T0177119

# WHO GIVES A GIGABYTE?

## A Survival Guide for the Technologically Perplexed

### Gary Stix and Miriam Lacob

John Wiley & Sons, Inc.

New York • Chichester • Weinheim • Brisbane • Singapore • Toronto

To our children, Benjamin and Madeleine,
in the hopes that the wise use of technology can enable
the development of a future world in tune with
the environment and all of its inhabitants.

This book is printed on acid-free paper. ∞

Copyright © 1999 by Gary Stix and Miriam Lacob. All rights reserved.

Published by John Wiley & Sons, Inc.
Published simultaneously in Canada

Figure 1.1 (page 10): courtesy of Texas Instruments. Figure 1.2 (page 11): courtesy of Lucent Technologies. Figure 1.3 (page 16): courtesy of Jared Schneidman Design. Figure 1.5 (page 20): courtesy of Digital Equipment Corp. Figure 1.6 (page 25): courtesy of Optitek Inc. Figure 2.3 (page 45): courtesy of the Massachusetts Institute of Technology Media Lab. Figure 2.4 (page 50): courtesy of Visible Productions. Figure 2.5 (page 51): courtesy of NASA. Figure 3.2 (page 59): courtesy of Lucent Technologies. Figure 3.5 (page 65): © John McGrail. Figure 3.9 (page 82): courtesy of the Massachusetts Institute of Technology Media Lab. Figure 4.1(page 91): courtesy of Northwestern State University. Figures 4.2 and 4.3 (pages 100–101): courtesy of Lasemaster. Figure 5.5 (page 137): illustration by Slim Films. Figures 8.2 and 8.3 (page 215–216): courtesy of Sandia National Laboratories. Figure 9.2 (page 251): courtesy of George Retseck. Figure 10.2 (page 278): courtesy of Ford Motor Company. Figure 10.3 (page 283): courtesy of Columbia University.

No part of this publication may be reproduced, stored in a retrieval system or transmitted in any form or by any means, electronic, mechanical, photocopying, recording, scanning or otherwise, except as permitted under Sections 107 or 108 of the 1976 United States Copyright Act, without either the prior written permission of the Publisher, or authorization through payment of the appropriate per-copy fee to the Copyright Clearance Center, 222 Rosewood Drive, Danvers, MA 01923, (978) 750-8400, fax (978) 750-4744. Requests to the Publisher for permission should be addressed to the Permissions Department, John Wiley & Sons, Inc., 605 Third Avenue, New York, NY 10158-0012, (212) 850-6011, fax (212) 850-6008, e-mail: PERMREQ @ WILEY.COM.

This publication is designed to provide accurate and authoritative information in regard to the subject matter covered. It is sold with the understanding that the publisher is not engaged in rendering professional services. If professional advice or other expert assistance is required, the services of a competent professional person should be sought.

*Library of Congress Cataloging-in-Publication Data:*

Stix, Gary.
    Who gives a gigabyte? : a survival guide for the technologically
perplexed / Gary Stix and Miriam Lacob.
        p.   cm.
    Includes index.
    ISBN 0-471-16293-0 (cloth : alk. paper)
    1. Technological innovations. 2. Technological literacy.
I. Lacob, Miriam.   II. Title.
T173.8.S75   1999
600—dc21                                                    98-35327

Printed in the United States of America

10  9  8  7  6  5  4  3  2  1

# Contents

Acknowledgments      iv

Introduction: Microchips and the Millennium      1

1   Computers 101: From Bits to Gigabytes and Beyond   9

2   Software: Making a Computer Bend to Your Will      33

3   Wiring the World: Telecommunications and
    Data Networks      53

4   Lasers: The Light Fantastic      89

5   All in the Genes: DNA Becomes an Industry      107

6   Medicine and Molecules: New Approaches to
    Drug Development      149

7   Spare Parts and High-Tech Flashlights:
    Repair Kits and Diagnostics for the Human Body      165

8   Material Improvements: Better Living through
    Advanced Chemistry      191

9   Mother Earth, Wind, and Fire: Energy for a
    Small Planet      223

10   Clean Machines: Technology and the
     Environment      255

Conclusion: Great Expectations      285

Further Reading      291

Index      295

# Acknowledgments

Any book is a collaboration beyond the efforts of the individual authors. This one is no exception. It would not have been possible without the abundant resources available to us from *Scientific American* magazine. A large measure of thanks must also go to Emily Loose, the John Wiley editor who helped enormously with the conception of the book. In addition, invaluable research assistance came from Robert Cox. A long list of manuscript reviewers helped correct and improve this volume. They include, in alphabetical order: Federico Capasso, Donna Cunningham, James Ellenbogen, Jerrold Gold, Paul Hoffman, John Horgan, G. Dan Hutcheson, Wayne Knox, John MacChesney, Mahmoud Naghshineh, George Musser, Joseph Pelton, Dr. Charles Pelizarri, Monica Roth, Phil Rotheim, Ricki Rusting, David Schneider, Pamela Tames, Chris Stix, Paul Wallich, Richard Wright, and Philip Yam. Gratitude goes to Edward Bell for obtaining illustrations, and to our children, Benjamin and Madeleine, for being so patient.

# Introduction

# Microchips and the Millennium

The watchmaker who lives in our neighborhood retired a few years ago. A German immigrant, he studied his trade for decades, becoming an expert in the finely balanced, exquisitely intricate machinery of Swiss clockwork. He decided to retire as the interlocking gears and cogs inside the timepiece started to be displaced by tiny blocks of silicon without moving parts. "I was an apprentice in Germany for ten years," he used to say. "I don't understand these new things. There is nothing to fix."

The days of the mechanically oriented watchmaker—or even the child tinkerer who pries open an appliance to see how it works—have faded. In their place is a flat, featureless landscape populated by countless electronic circuits. Its topography can only be inspected with powerful microscopes and its description rendered in the obtuse argot of DRAMS, ASICs, and SRAMs.

Microcircuitry surrounds us, not only in watch mechanisms, but in toasters, refrigerators, automobiles, and microwave ovens. The number of transistors in a typical American home outnumbers by far the nails attached to the wooden studs holding up its walls. The average homeowner has driven in a few nails, but has not a clue about the electronic viscera of the digital coffeemaker.

Thus it is with much of the high technology that saturates the modern world: even as the products of applied physics, chemistry, and biology disperse ever more widely, their workings become more obtuse to the

1

consumer who has come to depend upon them. The enabling technologies for the Nintendo game, the gene test, and the pocket cell phone require intimate knowledge of the deepest recesses of the atom, the living cell, and the electromagnetic spectrum.

To create sub-micron-sized electronic circuits that obey the orders of our relatively gigantic fingertips, chip designers tap into precise atomic knowledge of the material and electrical properties of such elements as silicon, germanium, and copper. With ever-more-refined tools and techniques, the genetic engineer manipulates a DNA molecule that would be almost a meter long if unwound from the nucleus of a human cell, yet is so infinitesimally thin that 5 million strands can fit through the eye of a needle.

As anyone knows who reheats a cup of tepid coffee in the microwave, it is entirely possible to enjoy the fruits of this technological age without having any idea of the functioning of the gizmo that excites the water molecules in the cup's recesses. True, many people wrestle with programming the VCR. Yet the majority of new technologies pay lip service to the credo of ease of use. Electronic mail speeds on its way by a simple click of a Send button. The many layers of computer programming that re-create a visual screen-based rendition of an office desktop remain opaque to most of us.

Maybe the inner workings of home appliances and the genetic code that shapes our destiny are best left to the experts. But the extent to which vast numbers of otherwise highly educated people remain technological illiterates periodically surfaces as a matter of societal and even personal concern. Arthur Koestler, a social critic, once described as "urban barbarians" people who own devices they do not understand. Koestler noted that the traditionally humanities-educated Western man will quite cheerfully confess that he does not know how his radio or heating system works, while he would be reluctant to admit that he did not understand a famous painting.

## What's in It for You?

There is a list of reasons for trying to educate oneself about the full complement of advanced technologies that drive modern industrial society. The foremost motivation is a selfish one. The creation of smaller and cheaper microcircuits and the diciphering of the genetic code are among the most exciting and stimulating challenges in science and engineering. In short, they hold an innate interest to the intellectually curious. One can follow the travails of solid-state physicists and engineers as they borrow hand-me-down methods from ancient lithographers to pattern circuit lines on a silicon chip that are less than one hundredth the width of a human hair. On a more basic level, will the breakneck pace of development of faster and cheaper chips continue? Or will technologists stumble on physical limits that will force designers to make radical changes in the way they build chips? Those hurdles could threaten the inexorable decline in the cost of microcircuits, which will have seen the price of an electronic memory circuit drop from $10 in the 1950s to an estimated hundred thousandth of a cent somewhere around the turn of the millennium.

Similarly, one of the few remaining "big science" projects—the Human Genome Project—faces equally daunting challenges. In the next few years it plans to elicit the codes for the roughly 80,000 genes for the stuff from which we are made, the proteins that constitute everything from brain cells to toenails. But unraveling the human genetic code will not be enough. For decades afterward, scientists will labor to determine, perhaps with only limited success, what the genes are used for. Behavioral traits that explain why some are optimists while others resemble Pooh's friend Eeyore may be determined by the complex interactions of different genes with one another and with a host of environmental influences. But the difficulties of understanding the effects of genes, or of nature versus nurture, may make human character ultimately unfathomable.

Another compelling rationale for learning as much as possible about things technical is that this knowledge can serve as a membership card to a societal elite fluent in such concepts as DRAMs and base pairs. Political and financial institutions are increasingly dominated by the technically and scientifically savvy. In the highly competitive business world, financial success frequently hinges on technological leadership.

Of course, this book will not furnish you with a specialist's credential, which is conferred only after the years of toil that result in an advanced degree in engineering or the sciences. But it will provide a gateway of sorts into the major disciplines—computers, biotechnology, materials science, and others—that have become the bulwark of the postindustrial society.

Becoming a technological heathen remains a danger, even for the specialist. The expert in one field may retain only a passing awareness of what occurs down the hall or at the company across the street. The pace of change makes it difficult to eschew the feeling of being the outsider. Fiddling with human genes to produce therapeutic proteins had the aura of science fiction thirty years ago, a period that falls within the life span of nearly half of all Americans.

In the new millennium, success as a manager or a project leader may depend on a multifaceted understanding of the interrelationships among technologies—how the computers made possible by novel material designs may process enough information to engineer new drugs from thousands or even millions of molecular candidates with slightly varying chemical properties.

Technological literacy also enables us to become better citizens. A knowledgeable public means that decisions on genetic testing, filtering of Internet content for minors, or placement of toxic waste sites will not be made by only a few business leaders, government officials, scientists, and engineers who understand and control these technologies. It does not take deep insight to realize that technologies are not just tools to make our lives richer and more convenient, but powerful forces that have a substantial impact on the whole planet. A basic understanding of the scope and limitations of new technology makes it easier for us to grasp related social issues free from hysteria and misconceptions. Protection of the diversity of species can be justified on ethical and scientific grounds. But it is also a potential source of genetic richness that may benefit agricultural and medical biotechnology.

We need to understand the power of technology as well as its limitations. Despite our ability to simulate life as computer-based representations, we are still creatures of the natural world, dependent on energy cycles and food chains. Carbon-based industrial emissions may cause catastrophic warming of the Earth's atmosphere. Here technology can assume a legitimate role, separated from its Frankenstein image. The prize could be nonpolluting energy generation and transportation and genetically engineered crops that can feed the world's billions without the water-table depletion and pollution caused by current intensive farming techniques. Advances in materials science hold the promise of manufacturing lighter and stronger materials that do not generate toxic wastes.

## From Computer Hardware to Bioremediation

The word *technology* is increasingly associated with computers and communications equipment manufacturers. The computer section of a well-stocked bookstore is filled with stack after stack of overthick volumes that explain the ins and outs of Microsoft's Windows operating system. This book is not geared toward the reader who wants to learn about setting up a personal Web site or mastering the intricacies of a programming language. Rather, it attempts to provide a concise survey of the workings of some of the most important advanced technologies.

The range of technology—from paper clips to big-screen televisions to nuclear power plants—is enormous. So we have tried to select the machines and processes that may most affect our lives today and beyond the dawn of the new millennium—and to explain the underlying physical, chemical, and biological concepts that make them a reality.

Chapters on computer hardware, software, and communications come first, not only because of the importance of these technologies for the ordinary citizen, but also because of their influence on every other engineering endeavor. The Human Genome Project, which is intended to decode all human genes, would not be possible without high-powered processors and database software. Digital technologies exert the same leverage in our age that steam power did in the early years of the Industrial Revolution. Understanding computers may constitute the foundation of technical literacy in the closing years of the twentieth century. "All technologies have consequences," writes computer scientist Gregory E. Rawlings, in a meditation on the seductions of computer technology. "But because our brain is the source of all our technology, and because computers let our brains think better and faster and cheaper, they have consequences all out of proportion to those of most technologies. Take computers away and the pace of biotechnological change would slow to a crawl. As would the pace of change in every other technical field."

The first chapter examines the basics of computer hardware logic, explores how microcircuits are manufactured, and forecasts future directions in computer hardware that move beyond conventional electronics. Our look at software details the process whereby ever more complex programming has made software simpler to use. It ends by chronicling the still-unsuccessful attempts to make software function more like the human mind. A chapter on telecommunications and data networks chronicles how millions of computers, distributed from San Jose to Kuala Lumpur, can add up to more than the sum of their parts. The Network, with a capital N, dictates that a single computer becomes just one neuron in a vast global megasystem for exchanging numbers, words, pictures, and speech.

Chapter 4, "Lasers: The Light Fantastic," describes the centerpiece of both futuristic warfare scenarios and the mundane bar-code scanner. The laser epitomizes the union between advanced scientific knowledge and sophisticated engineering. Beginning with its quantum physical principles, we proceed to the manifold uses of this artifact of James Bond movies.

Genetic engineering, the subject of chapter 5, has been called the discipline that will produce the milestones of the new millennium, rivaling the accomplishments of computers in the twentieth century. Ever since the 1953 discovery of the structure of DNA by the Nobel Prize–winning biochemists James Watson and Francis Crick, researchers have been probing and manipulating this helical-shaped molecule. Their discoveries have profound implications for human life and commerce. There are now techniques for cutting, pasting, and copying pieces of DNA. Deciphering the codes of life has fostered burgeoning industries that encompass agriculture, medicine, and waste management. Issues that we cover range from the Human Genome Project, which aims to spell out every base letter of our genetic inheritance, to the implications of altering the genes of important agricultural products like soybeans, corn, and cotton.

In chapter 6 we describe how growing knowledge of molecular biology—the biochemical basis of natural processes—drives new methods of drug development. A section on new directions in cancer treatment describes the evolution of new strategies for attacking malignant cancer cells without producing the damaging side effects of traditional treatments.

Despite the revolutionary impact of genetic engineering, the health sciences remain more than just gene-fiddling. The ability to see inside the body has become routine with such advanced diagnostic and imaging methods as magnetic resonance imaging and positron emission tomography. Chapter 7, "Spare Parts and High-Tech Flashlights," also highlights other areas in which science and technology play an important role in modern medical practice, including telemedicine and tissue engineering, a new field that holds the promise of developing replacement body parts.

The standard-issue personal computer is just one item brought to you through the insights of materials science, the multidisciplinary field outlined in chapter 8. Materials scientists meld chemistry, solid-state physics, and metallurgy to investigate the properties of solids. Their discoveries turn up not just in microchips, but in snack-food packaging, golf clubs, new bicycle models, and cooking utensils. "Intelligent" materials, incorporating sensors that respond and adjust to temperature or excessive vibrations, have begun to find their way into exercise equipment and downhill skis. In the future, material advances may be built right into the infrastructure of buildings, so that they can signal areas of stress and wear, and even possibly repair themselves.

The ability to extract and shape exotic materials depends on propitious energy usage and a benign environmental impact—topics explored in chapters 9 and 10. Without cheap, efficient energy, technical advance remains stultified. The industrial world's insatiable appetites may translate into overheating of the atmosphere, which may only be rectified through radical alterations to the status quo. Debates about global warming tie intimately to arguments about the availability of nonpolluting energy sources. The environmental sciences will help to examine the way that industry organizes itself. The nascent field of industrial ecology foresees waste products of one industry becoming the feedstock of another. A further trend toward "green" technology examines the retooling of production processes to use more environmentally benign chemicals and to minimize waste.

We conclude with a glance toward the forces that both foster and inhibit the pace of technological innovation—how science translates into consumer implements, how government can help or hurt, and how expectations for technological utopia often founder because of cost, practicality, and the gap between inventors' dreams and the limits imposed by the laws of physics. Throughout the text, timelines and glossaries are provided where we considered them to be important.

This volume is intended as both a survey of important technologies and a simple reference. It is not by any means a comprehensive tour. The waning of the Cold War has meant that civilian technology now outstrips military research and development in importance. So munitions—and related space technologies—have largely been left aside here. Advanced transportation is covered as it relates to how it may help the environment and produce energy savings. But some superstars of yesteryear, such as magnetically levitated trains, have been ignored because their prospects appear to have waned, given cost constraints, practicality, and the emergence of alternatives like high-speed conventional trains.

This book, of course, can be nothing more than a beginning. To those who wish to probe further, suggested readings, including many that we found helpful in our research, follow at the end of the book. We have omitted World Wide Web links because of their transient nature, but any key word or phrase in the text—*polymerase chain reaction*, for instance— should yield a wealth of material when entered in a Web-based search engine.

# 1

# Computers 101: From Bits to Gigabytes and Beyond

One hundred years ago, visionaries predicted flying machines and robots. But no one foresaw a machine with the capabilities of the digital computer. General-purpose processors have taken on the labors of legions of clerks, telephone switchboard operators, and bank tellers. They act as conduits to global communications networks, calculate household budgets, play chess, and even pilot planes. In 1900, silicon—the second most abundant element on earth—was the stuff of sand castles and glass windows. Now silicon chips contain millions of tiny circuits that constitute the underpinnings for a nearly $900 billion world electronics market, and the United States, in a year, will spend more on its information systems than any other part of its industrial infrastructure.

In 1955, when Microsoft founder Bill Gates was born, there were fewer than five hundred computers in the world. Now a single law firm in New York may have that many in one office, each with more computational punch than anything built in 1955. The world computer population today easily tops 100 million. Some pundits claim that more information is stored electronically around the world than is on paper, and that the amount of digital information will double every two years. And miniature computing devices are found everywhere, with dozens in

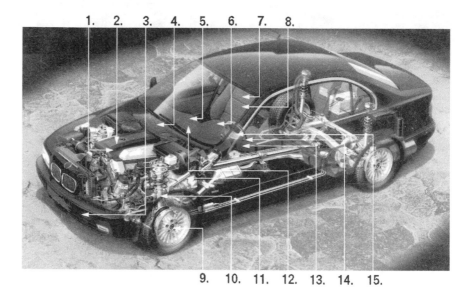

**Figure 1.1** We are surrounded by microcircuits. Microprocessor controls in a late-model automobile include (1) powertrain, (2) throttle, (3) cruise control, (4) body control, (5) sound system, (6) occupant sensor, (7) seat control, (8) display, (9) antilock brakes, (10) suspension, (11) power steering, (12) climate control, (13) security system, (14) air bag, and (15) instruments.

any home, and about fifty in a new car. Although the price tag for a modern chip fabrication facility can be over $1 billion dollars, chip designers have attained such successful economies of scale that microprocessors have become cheap commodities; they can be found in musical greeting cards and in five-dollar digital watches that can not only tell you the date and time, but may even be equipped with a stopwatch and a calculator.

Where did these powerful information-crunchers come from? While Jules Verne's futurist visions made no reference to computers, the technology did not emerge from a vacuum, either. In the seventeenth century, the French philosopher Blaise Pascal built an adding machine using gears, but found few takers when clerks became scared the device would put them out of work. In the late eighteenth century, Joseph-Marie Jacquard used punch cards to produce fabrics with exactly repeated patterns. In the nineteenth century, Charles Babbage, widely considered one of the progenitors of modern computer science, devised a five-by-ten-foot mechanical computer. But his "analytical engine" was so intolerant of error in the machining of its parts that he never got it to work.

Actually, vacuum tubes, invented one hundred years ago, became the basis of the first electronic computers. It took the crisis of World War II to prod their development. Battered by air bombardments and U-boats, the

British government built a computing machine called Colossus to crack the code of secret German communications, while in the United States, the 18,000-tube ENIAC was built with army aid to help plot bomb trajectories. Though completed after the war, ENIAC's early assignments remained military in nature; it was used to perform calculations to study the feasibility of the hydrogen bomb. Similarly, the development of integrated circuits, which started the process of installing millions of tiny electrical circuits into microchips, may never have taken off without the Cold War. Engineers developed such circuits—in which all of the formerly separate transistors were placed in one silicon chip—in order to shrink whole guidance systems into space capsules and missile warheads. This provided a ready market for them before they were really commercially viable.

Where in design and technology is computer hardware now, and where is it headed? Without any new technological breakthroughs, the

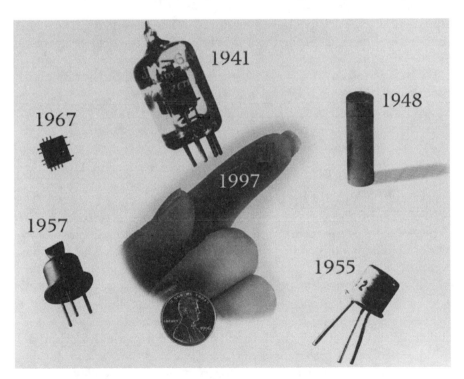

**Figure 1.2** Transistors have changed dramatically since their invention in 1947. Contrast the circa-1941 vacuum tube with a vintage 1997 Lucent Technologies digital signal processor chip containing 5 million transistors. Also pictured are the point-contact transistor, introduced in 1948; a 1955 transistor used in network communications; a signal amplifier built in 1957; and a microchip used in 1967 to produce tones in a telephone. This latter chip contains two transistors.

microprocessor—the thumbnail-sized chip that is at the heart of a computer—will soon contain tens of millions of transistors, packing more and more power and making the computer increasingly faster and easier to use. Soon users may talk to their computers routinely, not just poke at them with keyboards and touch-screens, and the computers themselves may well make current notebook computers seem bulky. Homes may be equipped with devices that control appliances, alarms, light switches, central air-conditioning, and heating. At the office, workstations may be seamlessly tied together with communication links hundreds of times faster than today's connections while, in turn, the boss may have limitless invisible access to our desktop computers. But despite these transformations, many of the basic principles by which all computers have worked since the days of ENIAC remain intact. And the way chips have been manufactured for decades remains virtually unchanged, thereby creating one of the principal conundrums for future generations of chip designers. The fabrication processes that fashion ever-smaller circuitry on silicon chips appear to be approaching their limits. When they do finally exhaust themselves sometime during the first few decades of the next century, scientists must find ways to exploit radically new and unproven designs of computers whose working parts are measured in billionths of a meter (nanometers). Research laboratories continue to labor on novel computational approaches that exploit some of the more bizarre properties of quantum mechanics—and that use biological or organic molecules to perform calculations (see page 26).

## An On/Off World: The Basics of Binary Logic

Whether it takes the form of a microprocessor embedded in the fuel injection system of a motorcar, or whether it is the latest personal computer equipped with power graphics, the digital computer has been based on two fundamental insights. First, designers had to understand how to break down and rewrite complicated problems in a way that allowed them to be solved in small, simple, clear steps that a machine could follow. The second insight was realizing that virtually all of math, at least arithmetic, could be performed with simple electronic switches that "knew" nothing except whether other switches were open or closed. This allowed information to be represented as a sequence of zeros and ones, a two-valued or binary code, which is the open or closed state of a single switch. These off and on states are known as binary digits, or bits.

At its most basic, a typical computer circuit is like an electrical circuit with a drawbridge that remains open unless a force pulls it closed. In a primitive computer, the bridge is a piece of metal that only lowers to a closed position to complete a circuit connection between two copper wires when a small electric magnet underneath tugs it closed. When this magnet, which is created with coiled wire, has a current running through it and therefore is "on," it pulls the bridge down to allow electrons to pass and complete a second circuit. Thus, if the magnetic circuit is on, the bridge circuit is on also. And if the magnetic circuit is "off," the bridge circuit is off. One circuit can "tell" what the other circuit is doing. Bridges that stay open after the magnet is on, or are closed until a magnet pops them open, are also easily designed. One could construct a sophisticated computer using bridges like this, but it would be ridiculously slow, take up as much space as a New York City skyscraper, and require miles of wiring. These kinds of bridges, called relays, were used in the 1930s to build the first electronic computers. In a more modern computer—like the common personal computer—the bridgelike switches are replaced by transistors, which are also electrically operated switches.

The trick in making a computer is to arrange these switches in a way that their being open or closed expresses some logical condition—denoting whether something is true or false. Computers break all problems down into simple "yes" or "no" decisions, which are expressed in a notation called Boolean algebra. Boolean algebra is named after the nineteenth-century mathematician George Boole. In Boolean algebra, certain words like NOT or AND describe given logical operations, which are similar to the way we make simple decisions. Computer scientists arrange a small group of switches, called logic gates, so that they will perform NOT, OR, and AND, plus other logical operations. An AND-type logic gate will only complete a circuit (or initiate an action) if switches A and B are closed. An analogy to this logical operation in everyday life might be, for example, if it is seven-thirty in the morning AND it is a weekday, then it is time to go to work: both conditions may be needed to complete the act of getting up. Both conditions need to be true to ensure that the action of going to work is initiated. Ultimately, every arithmetic and logical calculation that even the fastest, most powerful computer can accomplish is performed by as few as two logical operations.

How can big, complicated expressions be broken down into the physical actualities of wires either carrying current or not carrying current, the only two conditions that switches can handle? The answer is that everything is translated into binary numbers, or numbers expressed in "base 2," using only two numbers—zero and one—to express any value. (By contrast, our everyday number system uses base 10, comprising digits 0 through 9.) For example, 0 expressed in binary numbers is 0; 1 is 1; 2 is 10 (just as 10 comes after nine in base 10), and 3 is 11. For the convenience of the circuit designer, not only do binary numbers represent amounts as a series of on or off states, but all mathematical operations in binary are very simple, if tedious. The switches can easily open or close appropriately, though it takes a lot of them to come up with all possible answers—at least eight switches for every digit.

A binary computer also has another feature. If it can add two numbers, then it can also perform the other arithmetic operations: multiplication, subtraction, and division. Multiplication, for instance, can be carried out by repeatedly adding numbers, while a counter keeps track of how many times a number is added to itself. Subtraction can be performed using a "trick" in binary calculations called the *twos complement*, and division is the subtraction process done over and over. These are simplified examples of how they perform arithmetic operations. In practice, most modern computers have devised ways of adding and dividing that perform the same function in more sophisticated ways.

Thus a computer is basically a machine that adds and subtracts numbers. Those numbers represent an enormous variety of different things, from data locations in the memory to millions of points of color on a screen, to the words and typeface of these sentences. And the driving force of the digital revolution has been the growing ability of computer scientists to translate an increasing range of phenomena, such as sounds and video images, into binary code. Of course, it may take millions of these on-off switches to produce the bouncing ball of a child's computer game. And the wiring and programming between the switch and the ball is multilayered and extremely complicated. But beneath all the layering of a modern computer is the same "analytical engine" envisioned by Charles Babbage more than one hundred years ago. (Intrinsic to the functioning of hardware is computer software, which will be discussed in chapter 2.)

## Getting Small and Cheap: From Vacuum Tubes to Integrated Circuits

A computer constructed using relays would chatter like a loom and would probably have to be as big as a dining room table just to add nine and five. (That operation alone would require about fifty switches.) The first really workable switch that computer pioneers came up with was the vacuum tube. These could be switched on and off thousands of times a second, but they were big and could burn out easily, just like a lightbulb. Computers large enough to do meaningful work took up whole buildings, not just rooms, and a great deal of time was spent, in essence, replacing lightbulbs.

Following World War II, scientists came up with a much smaller and reliable switch that also became much cheaper. They were experimenting with semiconductors—substances that were neither good conductors of electricity nor good insulators. Common semiconductors included germanium, selenium, gallium arsenide, and silicon. Those scientists learned that the conductive properties of semiconductors could be altered dramatically by adding certain impurities, a process called "doping." By creating layers of what was called *n-type* (for negatively charged) and *p-type* (for positively charged) semiconductors, they found that they could create pill-sized devices that did everything a vacuum tube could do—amplifying electronic currents and switching them on and off. They had invented the transistor. Transistors were immediately hailed as the vacuum tube's logical successor, though during the 1950s they were expensive, and the sheer number of transistors required to make a computer turned it into a tangle of interconnecting wires. Engineers could conceive of circuits far more complicated than it was physically possible to construct using transistors. By the end of the 1950s, electrical engineers were up against what was called the "numbers problem": Too many transistors and little wires had to be connected by hand to create a higher-performance computer.

The solution, one of the most important technological developments in our century, was to realize that numerous transistors could be placed on a single piece of semiconductor. Specified portions of the chip could be turned into tiny electronic components—microscopic capacitors, resistors, and transistors—and then wired together by "photographing" or printing fine metal lines on the surface of a chip. The integrated circuit was born, and with it the last technological requirement for the computers we have today.

## FROM ANCIENT GREECE TO SILICON VALLEY: THE FABRICATION OF INTEGRATED CIRCUITS

Chip manufacturers use a technique derived from the ancient art of lithography to fabricate integrated circuits. The basic process of making transistors involves covering layers of silicon with a thin layer of material on which a pattern can be laid on the surface, while the material underneath remains protected. A pattern is produced on the chip by exposing it to ultraviolet radiation that is projected through a photomask—a quartz plate on which circuit lines are traced. The ultraviolet light that has not been blocked by the opaque portions of the mask falls onto the resistant material consisting of a photosensitive polymer. The exposed areas react to the light and are then washed away with a solvent. The now unprotected silicon is subsequently etched away with a plasma of ions. Impurities, such as arsenic or boron, are then implanted into the silicon, which alters the electrical conductivity of the device. Other steps in this process lay down metal and insulators to link together transistors into a circuit. These steps can be performed many times to produce a multilevel structure with millions of individual transistors in a

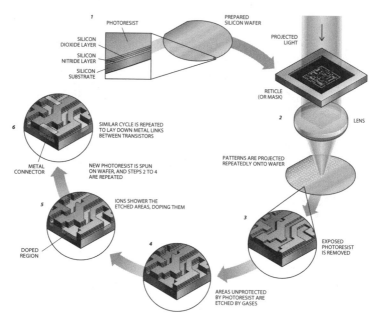

**Figure 1.3** Photolithography is used to build successive layers of material that make up a transistor. The cycle of steps can be carried out up to twenty times.

square centimeter. This method of fabricating electronic circuits is also being utilized to build microscopic mechanical machines that may in the future revolutionize materials science (see chapter 8).

The smallest dimensions of circuit elements printed onto today's most advanced microprocessor chips are about 0.25 of a micrometer (one millionth of a meter). But in their quest to build a gigachip that can pack a billion bits of circuitry into one square centimeter, manufacturers will have to produce circuit elements that are still thinner. A number of challenges must be overcome. One is to find a source of radiation with a short enough wavelength. A short wavelength is like a fine-tipped writing implement. It allows the patterning of thinner circuits that can pack more and more processing power on a chip. Another challenge is to decrease the distance between one chip wire and another without crashing up against problems of electrical resistance and errors. Most manufacturers use aluminum in their microscopic wires, but the electrical resistance of this metal is a barrier to making smaller wires. A new technology, the fruit of more than two decades of research at IBM and other institutions, uses copper wires, which are better conductors and can be as thin as 0.2 micrometers. To use copper in these dimensions, researchers had to develop a method of insulating the copper wiring so that its atoms did not contaminate the surrounding silicon and spoil the function of the transistors. For advanced chips, manufacturers must also develop lithographic technologies that can pattern circuit lines 0.10 of a micron wide or less. The ultraviolet light sources currently deployed will eventually have to be replaced by other sources of radiation, perhaps those using X rays or electron beams. In general, the development costs associated with designing new chips are enormous, and giant companies have formed cooperative research programs to spread the burden.

## How a Computer Works: The Basics of Memory and Processing

How is a computer different from the simplest calculator? An elementary calculator requires that a diet of numbers and commands be entered separately and slowly, one or a few at a time, from the keypad. A computer, on the other hand, has a memory that can hold sets of both values and instructions that are fed to the computer processor in a rapid, automated stream indicating what to to with particular values. It reads in amounts from a magnetically stored file, or the operator can type in a string of numbers before letting the machine work on them. A computer will store its instructions in advance in memory and then follow them one at a time. And outcomes in a calculation can affect which instruction the computer will then turn to: it can "jump" from here to there in the set of instructions called a program.

The real work is performed in the *central processing unit* or CPU of a computer. In the most basic calculator, values are entered into the processor directly. Computers, in contrast, pull numbers and instructions into the processor from "storage" outside the CPU and put new values back in, using the memory as a scratch pad. Essentially, computers boil down to memory and a central processing unit. All other parts—keyboard, monitor, disk drive, modem—are peripherals, devices for putting information into and pulling it out of the computer and providing permanent storage so it doesn't disappear when the computer is switched off. The disk drive, where information is stored permanently, is sometimes called secondary storage, while the memory is known as primary storage.

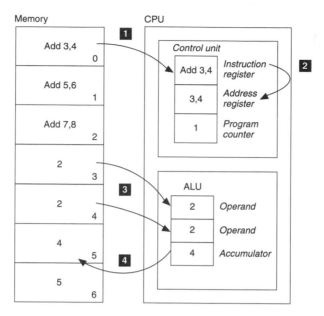

**Figure 1.4** When a computer performs an operation, the processor's control unit fetches the first instruction of the program from memory. The binary information is placed into a register (1). The instruction is decoded and the address register denotes where the relevant data are located in memory. Data for the instruction are retrieved and placed in the registers in the arithmetic logic unit (ALU) (3). The ALU carries out the instruction (2 + 2), places the results in the accumulator, then sends the results to memory (4). The program counter is then set to point to the next instruction in the sequence.

There are several vital parts to the portion of the central processor where calculations actually take place, all of which consist of circuits controlled by transistor on/off switches. When a computer must do something to a value, it brings instructions for processing data down through wires called a "bus" leading from memory into the processor, placing its binary information into a register in an area of the CPU called the control unit. Registers, the frontline trenches of the computer, are where values and instructions are stored temporarily. The control unit decodes an instruction and keeps track of what location in memory is to be accessed. The computer may next bring down other values from memory and place these bits in other registers. Then another important component called the arithmetic/logic unit (ALU) comes into play. The electric pulses sweep through the logic gates in the ALU, where values (which may represent letters or numerals) are added, subtracted, compared, or otherwise manipulated using the basic logic gates (AND, OR, NOT, etc.) and then some course of action is chosen. The process may be as simple as adding two and two. Or a comparison can determine whether a particular value is greater than, less than, or equal to another value. This is one of the critical elements of computing. If a number is greater than $2,500, for instance, then a credit limit may have been exceeded. After this operation is finished, the result is stored in a specific location in memory. Each step is accomplished in nanoseconds, and it may take many billions of these tiny steps to solve problems significant in the world. For this reason, there is another element in the CPU of any electronic computer that is essential to permit so many instructions to be synchronized accurately. This element is a vibrating crystal clock that controls the pulses of electricity that push the on/off information through the computer. Every pulse or tick sends a wave of electrons through all the switches or circuits in an orderly, controlled manner—pushing old values out of registers, herding new values in. In today's computers, pulses may be sent out as frequently as once every few billionths of a second, or nanoseconds.

## The Importance of Speed: Microchips and Moore's Law

The basic design of a computer—lots of switches in the CPU and memory—means a computer's speed will largely be determined by two factors in addition to clock speed. First, because even adding just two binary bits takes so many switches, how fast these switches can physically go from "on" to "off" and vice versa is critical. Second, because there is so much coming and going inside the CPU and between the CPU and memory, the longer the distance between any two points, the slower the computer. This puts a premium on getting all of the computer's parts very close together. With the invention of the integrated chip, transistors are etched and "wired" by the million, vastly reducing the cost per transistor. Now that they are much smaller and more densely packed, the travel time for electric impulses is greatly reduced. The feat of carving an entire central processing unit on one chip has created a machine that cannot only be afforded by the masses, but is so fast it can beat the big boxes of the 1960s at any computational game.

In the last quarter-century, microchip performance has improved 25,000-fold. Memory capacity has increased fourfold every three years. In 1964, Gordon Moore, a chip designer and a founder of Intel, made an off-the-cuff prediction that the number of transistors that could be squeezed into a single chip would double every eighteen months. This prediction was borne out, and "Moore's Law" has become an article of faith in the semiconductor industry. Performance is now expected to continue to double every eighteen months through the first years of the new millennium, with the technology we already have. In 1964, when Moore made his prediction, the most advanced chips contained sixty components. A few years later, the original Intel 4004 central-processor-on-a-chip, with 2,300 transistors, performed about as well as a mid-1940s-era computer with 18,000 vacuum tubes. A more recent PC chip of choice, the Pentium II, performs 233,000 times faster as it makes use of its 7.5 million transistors. Because each cycle of chips with twice the transistors of its predecessor has cost roughly the same, Moore's Law has also meant that the cost of computing power has also been cut in half every eighteen months. One computer analyst has calculated that if travel costs since the invention of the chip had dropped as dramatically, we could now fly to the moon for a dollar.

Chip performance has improved and will continue to improve as builders beef up underlying features of current designs. For example, the width of the registers, which determines the number of bits of information a computer can work with at a time, has repeatedly doubled over the

**Figure 1.5** The heart of the digital computer is the microprocessor. This 64-bit microprocessor contains 1.7 million transistors, and can process 300 million instructions in one second. The average personal computer contains more than 50 chips in addition to the microprocessor, including dynamic random-access memory chips (DRAM) that temporarily store instructions and chips that perform other functions, such as communications and synchronization.

years. The first CPU on a chip only had a four-bit register—it could only handle values up to 16. In 1975 the eight-bit register was introduced, and with it came the first micro-computer boom: such computers could now handle eight-bit units of information—one byte—and therefore a complete set of text characters. In 1981, sixteen-bit computers came into the fore with the introduction of the IBM PC. Now thirty-two-bit registers are standard, and PCs are "comfortable" handling values up to 2 to the power of 32, which comes out at over 4 billion. The pace of change continues. Sixty-four-bit registers are already being incorporated in advanced machines—and 128-bit registers are on their way.

When the register length is doubled, the power, and also the complexity, of a computer is squared. This includes the amount of memory it can locate and use. In addition to actual values of data, registers must hold and manipulate memory addresses that tell where data is stored. Longer registers mean that the computer can "count higher" and therefore specify more memory addresses. However, when more memory is added, the number of parallel electric paths running out of the registers (collectively called "the bus") will need to be doubled, so information may flow in and out without bottlenecks.

## Design Improvements: Pipelining and the Rise of RISC

For purposes of speed, designers want to keep the information that a user will be working with regularly as close to the registers as possible. This makes the computer work faster, since even electricity takes some time to go back and forth between memory and processor. To give the user faster access to data without any radical change in design, computer engineers add a special area of very fast memory called a "cache" as part of the CPU, a "holding pen" of transistors where values the registers most frequently call for are kept handy and close. Caches sometimes account for more than half of the components in a modern CPU, so important are they to speed. Other techniques for improving CPU speed include *pipelining*, whereby many pieces of information are moved through the processor at once, and *superscaling*, in which a processor executes more than a single instruction at a time.

With all these techniques in addition to cranking up clock speed, chip manufacturers have managed to double the power of chips about every eighteen months, fulfilling Moore's prophecy. In the beginning, much of the space on these chips was dedicated to the controller, the circuitry that translated an instruction into the electronic pulses that told the processor exactly what to do, generating the on/off information that actually controls the actions of the logic circuits and the registers. Often instructions had to be broken down into a whole series of pulses, a translation process that could take numerous machine cycles to complete. Enabling the computer to carry out these instructions gave it a big "vocabulary" of commands it could respond to.

In essence, computer scientists hard-wired as many bells and whistles as they could right into the CPU, an approach first taken in 1964 with the IBM 360. This approach, *complex instruction set computing* (CISC), dominated computer design into the 1980s. At that time, measurements showed that many of these functions were rarely called upon in practice. This gave rise to the idea of eliminating them. If this could be done, the computer could run a little program from memory (part of the big program being run) that would tell it how to handle the task. Among the advantages were that the machine would never have to wait while instructions were being broken down into numerous series of impulses. More important, moving instructions and values through the CPU would become faster, as information would move regularly through the processor with every clock cycle. The most significant effect, however, was that designers could make good use of the space freed up inside the chip. They could add more registers and hence keep values being worked on handy in the CPU instead of shuttling them back and forth between the CPU and memory or a cache.

Called *reduced instruction set computing*, or RISC, this approach has emerged as a significant competitor with CISC. Chips based on RISC premiered in the late 1980s in workstation computers. These could handle scientific problems and graphics beyond the capability of non-RISC-based PCs. The chips controlling most laser printers and copiers are RISC-based, and RISC architecture moved into a popular personal computer when the Macintosh started to be built around the Power PC chip in 1993, which has a RISC-based design.

## Parallel Processing: Speeding Up by Cutting Problems Down to Size

Another computer technology available today that greatly speeds up computation time is *parallel processing,* in which the idea is to break problems down and enter them into a number of processors working in unison. If eight processors are attacking some problem, the theory goes, the computer should get through it eight times faster than it would with only one processor.

Processing data can be analogous to trying to get a morning rush hour's worth of cars through a toll booth plaza. In an eight-bit computer, eight lanes of data pull in unison into the row of toll booths—the register—and then move on. Expanding the numbers of both lanes and toll booths increases the amount of cars a toll plaza can handle. A sixteen- or thirty-two-lane system will greatly enhance movement—one advantage of having wider registers. (RISC processing, with all its registers, tries to eliminate the commute altogether: the most important cars just stay inside their own booths, while processing all cars is kept quick and simple. Caches try to reduce the commute by keeping cars parked nearby.) Parallel processing doesn't widen the plaza. It relieves traffic by adding whole new systems of roads and tolls identical to the original one. It lines or stacks up identical toll plazas.

Constructing the hardware necessary to exploit parallel processing is the easy part, but feeding data into a series of yoked CPUs in an ordered, significant manner is enormously difficult. The superhighway of information leading to the plazas is meaningless unless it flows into all the booths in an ordered, controlled manner. The pattern of the cars as they come off the ramps and cloverleaves leading to the multiple-level bridge becomes enormously complicated. Similarly, problems must be broken down into controlled streams of information. This problem of coordinating so many "carloads" of data through a complicated network of parallel paths so that they can be worked on by many CPUs at once has created new software concerns. These have become the stumbling blocks for successful parallel processing. But parallel computing has become essential to such number-intensive fields as weather forecasting, oil exploration, and aerodynamics.

### Reprogrammable Chips: Customizing on the Fly

Another design innovation in recent years allows for more flexibility. Traditionally, computer engineers must decide early on in the design process whether a chip will be a specialist or a generalist. It can be tailored for a specific function—high speed, for instance—or it might resemble a Swiss Army knife that slogs through virtually any mathematical or logical operation encoded in a computer program. An alternative that has begun to emerge in recent years gives designers the best of both worlds, a chip that can be reprogrammed to the task at hand. These integrated circuits, known as *field programmable gate arrays* (FPGAs), can be altered at virtually any time during the life of the chip. Usually the logic circuitry of a chip cannot be changed after the chip is manufactured. On an FPGA, though, blocks of logic gates can be rewired in milliseconds by sending signals to the circuits on the chip. For instance, a chip for recognizing images can adjust its circuitry: switching between processing the image of a high-speed aircraft or focusing on the gait of an ambling person. Instead of incorporating separate chips to follow either slow- or fast-moving objects, a single chip could rewire itself as needed to perform both operations.

## Magnets, Holograms, and Molecules: Computer Memory and Storage

The output of increasingly powerful computer processors requires hard drives that store billions of bytes of data on their magnetically coated surfaces. With a hard drive, information and programs are on alert to be run, written to, or read from. Hard disks have metallic surfaces laced with magnetic filaments that are flipped into either "on" or "off" positions—up or down—by the magnetic fields generated by heads with electronic currents that run one way or the other, depending on whether they are writing a one or a zero. The read/write heads can retrieve information from a location in thousandths of a second. But this retrieval time is too slow to feed the computer's processor. So, while the program is running, it is copied from the disk into fast memory chips called *random access memory*—or RAM. This memory can be written into as well as read from. It stores applications programs when they are being run, along with the files that they are manipulating. Most of a personal computer's memory is called dynamic RAM or DRAM.

As computers get through data faster and faster, the speed at which the machine pulls information out becomes crucial to overall performance. Writing and reading information to and from storage disks—be they floppy or hard—causes a pause in the processing. CD-ROM disks may store 640 million bytes of information (300,000 pages of double-spaced text), but reading this data takes time. And some information-intensive institutions—hospitals and law firms—have had to resort to "jukebox" multiple-disk machines to keep all the information they need at their computer's fingertips.

## Holographic Storage: Information Retrieval by the Page

One method that may enhance both information density and speed of access is optical storage using multidimensional imagery called holograms. Whole "pages" containing perhaps a hundred thousand bits of information each may be read in a flash, with millions of pages accessible in thousandths of a second. Already, holographic storage devices exist to help computers sort through digital scans of fingerprints; such machines can compare a print to a thousand other prints in one second. A robotic car has been built that steers itself by comparing images it sees through a video camera with images stored in permanent memory, which it leafs through almost spontaneously. With holographic storage, information doesn't trickle into a computer through a few wires; it pours off pages in torrents.

*Hologram* means "whole picture." In regular photography, light-sensitive chemicals only record the intensity of light as it reflects off an object. A hologram stores much more: it records how an object diffuses and scatters light. When re-created, the hologram interacts with light exactly as does its three-dimensional counterpart. Holograms are created with a coherent laser beam—light waves of a single wavelength in which the crest and trough of each wave move in lockstep. The laser light is split in two. One beam, called the object beam, is reflected off the object, and in doing so will have its waves knocked out of phase and diffused. The second beam, called the reference beam, is then directed back onto the object beam so that they cross each other and in doing so create a complicated interference pattern. This is recorded on photoelectric sensitive film or a crystal. (See chapter 4 for more on lasers.)

When a beam of light identical to the reference beam is then passed over this stored pattern, it scatters light in a way that re-creates the image of the original object. For holographic data storage, three extra devices are added to this arrangement. The object that gets "holographed"—shot with the object beam—is a small screen made up of a matrix of electric diodes (what lights up a digital watch) that are either "on" or "off." This checkerboard pattern of on-and-off diodes represents tens of thousands of bits to be stored by the hologram. A crystal of photosensitive chemicals is

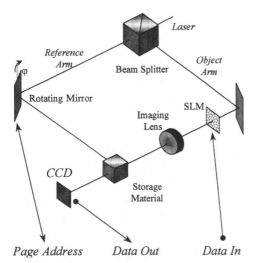

Figure 1.6 In one method that uses lasers to store and re-create data using a hologram, the data to be stored is formatted into a page of pixels in the object arm by a device called the *spatial light modulator* (SLM). The reference beam and the object beam interfere in the storage crystal to create a hologram. The angle of the reference beam is changed by the rotating mirror to record different pages. To read the page, the reference beam alone is directed to the same angle at which the data was recorded. An electronic camera, a charge-coupled device (CCD), then reads the desired page.

used to hold the resulting diffraction pattern. When the reference beam alone shines through this pattern to re-create the information, the hologram is beamed on a small device that reads the original pattern and sends out appropriate signals to show whether a square was on or off.

Holograms can be stored very compactly. Slight thousandths-of-a-degree changes to the mirror that directs the reference beam will "carve" a new hologram in a crystal without corrupting any hologram already stored. By the mid-1990s, experimenters could store ten thousand holograms in a single location. Sixteen locations could be mapped into a crystal, giving one crystal a capacity of over 10 billion bits. Access for locating and retrieving any page of that data was down to 40 billionths of a second. Although problems remain before holographic storage enters the consumer marketplace, the basic technology has been proven.

## Atomic Force Microscopes—A Terabit in Your Pocket

A competitor with holograms in the advanced storage research realm is the *scanning probe microscope*, an instrument with a needlelike tip that can pick up and move individual atoms or molecules. Gerd Binnig and Heinrich Rohrer received the Nobel Prize for Physics in 1986 for the invention of one of these devices, called the *scanning tunneling microscope*, which is now routinely used in basic scientific research. Binnig and other IBM researchers have more recently shown that a related instrument, the atomic force microscope, can use its nanometer-scale metal tip to "write" a data bit on a spinning plastic disk the size of a penny. The heat from the tip causes a tiny indentation in the disk surface; it can also probe into each dent to sense where a bit resides. In theory, a 1.4-square-inch area, accessed with arrays of tips suspended from microscopic cantilevers, could store a terabit (1 trillion bits) of information. In a more distant scenario, the technology might presage data storage using a single atom or molecule to represent a digital one or zero, permitting storage of possibly a million gigabits per square inch. But moving single atoms currently requires that surfaces be chilled to an impractical temperature near absolute zero.

## Molecular Computing: Harnessing the Power of DNA

A feature on a chip may only be one three-thousandth of a human hair in width, but this is still thousands of atoms wide. Viewed from our perspective, these structures are miraculously small. But from the "perspective" of an atom, we still have a long way to go in terms of potential downsizing.

Because we are clumps of matter trillions of atoms big, we find it difficult to manipulate things that aren't also a huge number of atoms. We have to work in bulk. Biology, on the other hand, works effectively at the atomic level, binding things together molecule by molecule. If we could be as deft as nature at manipulating individual atoms, we probably could build computers far smaller and more powerful than anything we have today.

The idea of using biological agents to build molecular computers is gaining adherents. Computer scientists are currently intrigued with the computational power of DNA—or deoxyribonucleic acid—the long, spiraling molecules that store the genetic information of every living organism on earth (see chapter 5).

Our current magnetic storage systems can retain one bit of information on a thousandth of a cubic millimeter. DNA can store exponentially more. If so, why not use strings of DNA to replace the binary digits used in conventional computing? The information-carrying molecules in DNA, called base pairs, might be combined, copied, and deleted, just as can digital bits. Proponents of DNA computing point out that the DNA crammed into the nucleus of a human cell holds, in its billions of molecules, enough information to build any part of your body. The three-quarters of a pound of DNA inside you has, roughly speaking, the storage capacity of all the computer memory ever built.

The hard part for humans in tapping these vast resources of representational power is "reading" and "writing" the DNA: peering into DNA sequences and constructing and manipulating the molecules that encode information. Here computer scientists have turned to techniques developed by molecular biologists (see chapter 5), using enzymes to "write" to DNA, multiply molecules that encode correct answers, and "read" results. In essence, the computer hardware is reduced to a laboratory flask with a DNA solution that contains the enzymes needed for computational processing. The enzymes combine, copy, and extract the strands of DNA, each of which represents a certain value. By performing this operation repeatedly, this DNA machine ultimately arrives at a solution to a given problem.

What DNA computing has going for it, in addition to tiny size, is the fact that you can make trillions of DNA molecules at a time, and these molecules will go through billions of reactions at once. What will likely be considered the first DNA computer was a drop of DNA solution a fiftieth of a teaspoon in size that, in 1995, came up with the right answer to a classic route-connecting problem: How does one travel between seven different points without going over the same path? A computer scientist, Leonard M. Adleman, brewed up millions of copies of seven different strips of DNA, each with a unique twenty-unit code representing each "city." He then let these strips combine randomly into bigger strips, which would represent possible "itineraries" between the cities. Next he sorted the DNA looking for his answer, culling first for strips that were the right length, then looking for strips that had one and only one code representing a city in them. Because he had billions of molecules stewing in his drop, he was virtually assured of having at least one that matched the conditions of the solution. He made copies of these and "read" them using standard laboratory techniques.

There is no reasoning in DNA computing, beyond what the scientist can bring to it through culling and selectively reproducing molecules. What DNA computing offers is a massive number of tries at a problem. It's as if one were posed the problem of constructing a key for a particular lock. Working alone, the locksmith might try to measure the lock and take it apart in order to measure the depth of its cylinders. The DNA computer would take a different approach. Almost instantaneously it would throw together all the keys that could possibly be constructed. The job then becomes to sort through the keys and eliminate all that are too long, too short, or otherwise don't meet the conditions of the lock that needs to be opened.

Many real-world problems are best solved through this type of massive attack; most important among these are what mathematicians call satisfiability problems, which involve long logical equations with conditions that must be met. Scheduling, route, and encryption problems appear ideal for DNA computers; scientists have already worked out the steps needed to solve many of them. Manipulating the DNA, however, remains tedious. Dr. Adleman's simple experiment, for example, took seven days of lab work. So no one expects the next memory card to be a vial of DNA. But a number of scientists do contend that computing with DNA will be a practical possibility in the future.

DNA is not the only biological molecule that could become important in the world of computing. Pigments, produced by bacteria, that change configuration when exposed to light, may become "yes/no" switches in storage devices for optical computers—computers that send streams of photons instead of electrons through logic circuits. An optical molecular computer would have the same advantages as fiberoptic communications: lightning speeds and a huge capacity. There could also be exponential increases in the storage capacity of an optical memory disk. Memory components would be grown by microorganisms instead of manufactured in a factory. It has been estimated that a device based on pigment produced by bacteria could have a capacity three hundred times greater than CD-ROMs.

Nonbiological organic molecules are also vying for a role on the outer limits of computing. Organic molecules are just molecules based on atoms of the element carbon. Researchers have shaped chains of hexagonal benzene organic molecules into a chain, creating an almost unimaginably narrow "molecular wire"—that conducts electricity. And investigators are laboring on building multiple organic molecules into structures that act like molecular transistors, capable of amplifying the current as well as of switching its flow on and off. Still other groups are striving to design and test individual molecules that behave as memory cells, capturing and releasing single electrons to store ones and zeros. One team has designed an organic molecule that would add two binary numbers when a current is passed through it. The dimensions of organic molecules, which measure only a few nanometers across, mean that nearly a trillion trillion identical molecular circuit structures might be manufactured simultaneously, about a billion times as many circuits as there are transistors in the entire world.

## Quantum Computation: The Ultimate System

If we are learning how to record information on individual molecules, why can't we go that one step further and "write" onto individual atoms? The answer is, we might be able to, though when we get this small, nature makes the job difficult. At the atomic level, there are a range of energy states that would appear perfect for expressing binary information. The electron of a hydrogen atom, for example, can be in a low-energy state and then be nudged with light into a high-energy state. Alternately, the spin of particles in a magnetic field can be "flipped" from one direction to the other.

But there are two major complications: first, it is very difficult to target our relatively large tools to change the state of a single particle in one atom, and, second, when we work at this level, we discover that the components of matter no longer work logically and predictably. The classical and causal "laws" first discovered by Isaac Newton in the seventeenth century are supplanted by the esoteric rules of quantum physics. For scientists trying to build atom-sized computer components, the good news is that the energy levels of electrons and photons must come in distinct amounts or packets known as quanta. The bad news is these particles can do what appear to be "impossible" things, such as tunnel through matter and be in two places at once.

According to quantum physics, the precise location of an electron cannot be pinpointed—but only regions where it *may* be located. It can leak out of wires, and jump through the walls of "dots," submicron-sized boxes made of semiconductors that can hold an individual electron. In this world, yes and no become "probably yes" and "probably no," and can also represent "maybe."

According to quantum theory, atomic particles exist at discrete energy levels. If one thinks of electrons surrounding the nucleus of an atom in concentric orbits, each electron circles the nucleus at a specific energy level. Quantum particles—electrons and photons, for example—can also be energized to a point where they appear to be in two energy states at once. A hydrogen electron also can, with the right wavelength of light hitting it, enter a state called superposition—it can be in a "maybe" state somewhere in between its two prescribed energy levels. It is in essence in both energy levels until the moment it interacts with another particle. (There is no continuum between this and the next energy level, but literally a quantum leap to a different orbit.)

It would appear that this quantum ambiguity would undermine the ability to set an electron in an on/off state. In fact, the properties of particles in superposition potentially give quantum computers enormous advantages over their conventional brethren. These quantum computer bits, called *qubits*, introduce "parallelism" into the logic of the computer, since half the time they break into one energy state and half the time into another. Each energy state, which is achieved by the atom's absorbing or shedding a particle, represents a zero or a one. More important, such a system can get the job done more quickly. Qubits can represent both one and zero or yes and no at the same time, allowing a computer to explore all possible values at once, and all possible outcomes. Called *quantum parallelism*, this ability should allow quantum computers to be much faster at certain tasks than digital computers.

Computer scientists have already come up with lightning-speed algorithms (steps for solving a problem so specific that even a machine can follow them) for tedious chores such as factoring large numbers and searching through databases. Quantum computers should also excel at code-cracking and at providing models of subatomic particles—a boon, since the quantum laws these particles obey are awkward on a digital machine. The hard part is building the computer. Scientists must build reliable systems that allow for the fact that quantum computers will tend to produce errors, as particles switch from state to state seemingly without rhyme or reason. The biggest hurdle is that particles will remain in superposition—two or more places at once—only when untouched by stray electrons and photons. Any kind of nudge from an electron or photon on the qubits causes them to lose their superposition, at which point they convert to ones and zeros—scientists say they lose their "coherence." Thus, quantum computers are unlikely to become everyday workhorses parked on a messy desktop. Rather, these potentially powerful computers will do their work isolated in some as-yet-to-be-devised way from the world that will be dependent upon their output.

Nevertheless, researchers have progressed beyond mere theoretical musings. In the spring of 1998, scientists from various institutions, including the Massachusetts Institute of Technology and IBM, reported building a prototype quantum computer able to execute a parallel search for data. It found one out of four possible binary states in a single step, rather than through the two or more tries that would have been necessary if the qubits were not in superposition.

## Robotics: A Less Ambitious Future

In 1961 the first industrial robot started unloading parts that were to be used in a die-casting operation. Since then, about 500,000 robots have entered the "workforce," employed on assembly lines and in manufacturing plants worldwide for tasks that are exceedingly repetitive or dangerous. Although they have earned a lasting place in the factory, robots are not as omnipresent as originally expected. For the time being, robots—machines that respond to the environment as they manipulate objects—remain niche players. The reason is that building in "smarts" and the mechanisms to respond to those smarts is expensive. Managers have learned there is little reason to invest in a costly robot when a retooled traditional machine, with more human assistance, can perform the job equally well. Robotics still remains an active field of research in the ongoing attempts to develop artificial intelligence (see chapter 2).

---

### BITS, BUSES, AND CPUs:
### A GLOSSARY OF COMPUTER HARDWARE TERMS

**binary/base 2**   A numbering system that is the basis of all calculations in a digital computer. Base 2 uses ones and zeros. The place value of the digits is powers of two.

**Boolean algebra**   A system of logical algebra used in computer design to break down complicated problems into simple, logical relationships like AND, OR, NOR.

**bit**   Binary digit: the basic unit of binary information—a one or a zero.

**bus**   Rows of wiring that carry information from one part of the computer to another (or to different sections of the CPU). The number of strands in a bus usually matches the width of a computer's register so that values may be transported easily.

**byte**   Eight bits, or enough information to contain an encoded letter or a numeral.

**cache**   The memory area inside the CPU for storing frequently used values, to avoid the long trip to and from the RAM.

**central processing unit (CPU)**   The part of the computer that runs the machine and performs the calculations.

**clock speed**   The number of pulses of electricity that the central computer clock sends through the computer in a given time, usually expressed in terms of millions of cycles per second.

**complex instruction set computing (CISC)**   A design theory that tries to maximize the number of instructions a CPU can recognize and carry out, in an attempt to make the CPU more responsive.

**controller**   The part of the CPU that generates the electronic signal that directs the rest of the CPU.

**integrated circuit**   A single piece of semiconductor housing numerous electronic components—transistors, capacitors, and resistors.

**logic gate**   An arrangement of a small number of circuits that performs one logical operation.

**memory**   The part of the computer where information is kept at the ready for quick access. Instructions and values are stored and written to Random Access Memory (RAM). Startup instructions and information for the computer are usually kept in Read Only Memory (ROM), which may only be retrieved by the CPU—but not easily altered. Information in ROM is not lost by turning off the computer.

**parallel processing**   A technique to boost computer power by having more than one CPU work on the same problem in concert.

**pipelining**   Moving numerous pieces of information through a CPU at the same time to improve computer speed.

**photolithography**   A process in which electronic circuits are patterned and etched into a semiconductor integrated circuit.

**reduced instruction set computing (RISC)**   A computer design theory that tries to maximize speed by keeping the set of all instructions down to only those that can be performed in one clock cycle, easing pipelining and superscaling, and also opening up CPU space for more registers.

**register**   A row of transistors in the CPU that holds some value to be used in a calculation, or that is otherwise crucial to its work.

**semiconductor**   A material that conducts electricity better than an insulator, such as rubber, but worse than a conductor, such as copper. Its features allow the precise control of an electric current at very small scales.

**superscaling**   Enabling a CPU to carry out more than one instruction at a time.

**transistor**   A device fabricated from a semiconductor for altering electric currents. In computers, transistors are used mostly to allow one circuit to switch another circuit on or off.

# 2

# Software:
# Making a Computer
# Bend to Your Will

In the past fifty years there have been radical changes in how computers are used and who uses them. New building techniques and materials have led to the steady democratization of computer power. The first computers were used to probe guarded military secrets, cracking an enemy code, for instance. Their power was then gradually shared with society's corporate and scientific elite. Transistor-based machines allowed computers to become fixtures in universities and corporate accounting offices. Finally, inexpensive microcomputers broke through the remaining barrier between computational wizardry and the person on the street, who, armed with the personal computer, could use it to track the progress of a mutual fund, prepare a term paper, rewrite a résumé, or play a quick game of solitaire.

Personal computers are now run by anyone who can move a mouse, read a menu, and click on an icon. Their ease of use is possible not only because they have become so affordable, but because layer upon layer of programming obscures the inner workings of the machine from the casual user. We don't directly tell the machine what to do anymore; we now tell a software program what to do by making simple selections, and the program tells the machine what to do. The software relays our commands to the central processor in the same language of ones and zeros that was spoken by the first computer scientists.

Ironically, at the same time that computers have become ubiquitous, far fewer people exercise direct control over the hardware, or even the software. Before, to work the computer was to program the computer. Now, to work the computer is to count on programs written by a new priesthood, that of the software developer. And at this point the achievements of computer hardware engineers and software developers have become so interdependent that it is difficult to tell where the hard-wiring of the computer ends and the software begins.

On the hardware side, advances in chip design have produced a fingernail-sized microprocessor containing millions of electrical circuits, enabling the development of more powerful machines. Armed with this raw power, software developers have created increasingly sophisticated and "user-friendly" programs with gargantuan appetites for computer bits. The power and capabilities of these programs are a function of the creative intellectual capacities of the programmer. And software has become just as important as hardware. It is no surprise, then, that the richest man in the world, Bill Gates, is the founder of Microsoft, the leading purveyor of computer software. Many observers believe that the future of computing will be defined as much by changes in software as by developments in actual hardware.

## From Machine Code to Assemblers:
## The Progression to "Higher" Programming

Ever since computer scientists in the 1940s learned how tedious it was to tell the Harvard Mark 1 computer what to do by switching levers and cables, all computers have worked by pulling instructions out of memory. These instructions are placed one by one into the CPU register that runs the controller, which in turn sends out the electronic pulses needed to get the job done. The controller sends signals to another part of the CPU that will perform a specific action—add two numbers in two other registers, fetch some value from memory—as dictated by the instruction. The controller, in essence, has been built to "understand" and respond to values in the register.

The instructions that make the computer perform calculations are called *machine code*. Collectively they make up the instruction set of the computer. The computer can only perform tasks that can eventually be broken down into some series of these codes. At the beginning, there was only one way to program a computer, which was to feed it machine code—the groups of zeros and ones that its controller understands. Programming then was called "coding" because that is exactly what programmers did: they translated tasks into strings of ones and zeros that told the computer what operation to perform.

Feeding in long strings of machine code tested human patience and made for programs that were as difficult to read as Sumerian cuneiform. Programmers quickly decided their job would be much easier if they could move one step away from the binary language of the CPU. Instead of speaking in ones and zeros, they would use simple mnemonics such as "mov" and "add," with each word corresponding to a unique string of instructions in the lower-level machine language.

Special programs, called *assemblers*, translated these mnemonics into machine code. The "language" with its mnemonics fed into the assembler was called *assembly language*, and some programmers still write some code in assembly languages. The good news is that assembly language resembles machine code, which allows a programmer to tell the computer exactly what to do each step of the way. The bad news is that because you

```
           PRINT  NOGEN
PROG8      START  0
CARDFIL    DTFCD  DEVADDR=SYSRDR,RECFORM=FIXUNB,IOAREA1=CARDREC,C
                  TYPEFLE=INPUT,BLKSIZE=80,EOFADDR=FINISH
REPTFIL    DTFPR  DEVADDR=SYSLST,IOAREA1=PRNTREC,BLKSIZE=132
BEGIN      BALR   3,0             REGISTER 3 IS BASE REGISTER
           USING  *,3
           OPEN   CARDFIL,REPTFIL OPEN FILES
           MVC    PRNTREC,SPACES  MOVE SPACES TO OUTPUT RECORD
READLOOP   GET    CARDFIL         READ A RECORD
           MVC    OFIRST,IFIRST   MOVE ALL INPUT FIELDS
           MVC    OLAST,ILAST     TO OUTPUT RECORD FIELDS
           MVC    OADDR,IADDR
           MVC    OCITY,ICITY
           MVC    OSTATE,ISTATE
           MVC    OZIP,IZIP
           PUT    REPTFIL         WRITE THE RECORD
           B      READLOOP        BRANCH TO READ AGAIN
FINISH     CLOSE  CARDFIL,REPTFIL CLOSE FILES
           EOJ                    END OF JOB
CARDREC    DS     0CL80           DESCRIPTION OF INPUT RECORD
IFIRST     DS     CL10
ILAST      DS     CL10
IADDR      DS     CL30
ICITY      DS     CL20
ISTATE     DS     CL2
IZIP       DS     CL5
           DS     CL3
PRNTREC    DS     0CL132          DESCRIPTION OF OUTPUT RECORD
           DS     CL10
OLAST      DS     CL10
           DS     CL5
OFIRST     DS     CL10
           DS     CL15
OADDR      DS     CL30
           DS     CL15
OCITY      DS     CL20
           DS     CL5
OSTATE     DS     CL2
           DS     CL5
OZIP       DS     CL5
SPACES     DC     CL132' '
           END    BEGIN
```

**Figure 2.1** A page of instructions in assembly language. This example shows an IBM assembly program for reading and writing out a record. The left column contains symbolic addresses of various instructions or data. The second column lists the operation codes for the activity needed; the third column describes the data on which the instructions act; and the column on the far right is for comments about the activity on that line.

have to specify every step, it takes a large number of instructions to accomplish any significant task. What sounds like a simple request, such as "add two values," must be broken down into all the steps the computer must actually take.

The "higher" computer languages, such as BASIC, FORTRAN, and COBOL, were created to allow programmers to get more done by enabling them to use relatively humanlike phrases that translated into whole paragraphs of machine code. With a higher-level language, programmers write in sentence-like declarations that are initially incomprehensible to the computer. A special program called a compiler can translate the instructions written in a form people can understand to the ones and zeros of machine code. (Another kind of program-translating software, called an interpreter, converts instructions from higher-level language to machine code one line at a time rather than all at once. Interpreted programs must be retranslated each time they are run.)

## Enter the Operating System

Another aspect of early programming was that each programmer had to write machine instructions from scratch to perform such common tasks as telling the computer how to manipulate devices that communicated with the outside world (such as keyboards and printers), or how to store information on magnetic disks or tapes. This programming was difficult and error-prone.

A better way was to have the instructions for these tasks always at hand to be loaded into memory and to be called on as needed by a set of indispensable housekeeping programs. This software—which also starts up the computer—is called the *operating system*. Other programs, the applications software that ranges from spreadsheets to word processors, invoke little programs or routines from the operating system when they want to do any machine-specific task, such as free up space in memory, write to a hard disk, or display information on a screen.

Operating systems like MS DOS, which Bill Gates parlayed into his fortune, are the collection of programs that run the computer system. Creating an operating system or adding additional tools to its basic collection is called *systems programming*. All these tools have been created to empower what the more casual user thinks of when they think of the word "software"—the big commercial applications we use to write novels, compose spreadsheets, or keep track of our accounts receivable or our scores playing virtual golf.

## SO WHAT DO THOSE FUNNY NAMES AND ACRONYMS MEAN, ANYWAY?: A BRIEF LOOK AT COMPUTER LANGUAGES

In the 1950s, scientists wanted a language that would be relatively easy to learn and also one that would produce programs that could run on many different types of computers. Developed by IBM and released in 1954, the answer was the Formula Translating system, or FORTRAN, a language still in use. FORTRAN is rich in numerical operations but weak in its ability to process nonnumeric data such as strings (lines of text).

Business lacked a higher language appropriate to its needs until COBOL—Common Business Oriented Language—was developed. COBOL took off in 1959 when the U.S. government insisted that it be used by companies hoping to gain government contracts in the computer field. PL/1, or Programming Language 1, which was introduced by IBM in 1964, was designed to be a universal computer language, combining COBOL's better file- and data-handling capabilities with FORTRAN's numerical abilities. BASIC was developed a year later at Dartmouth College to be an easy-to-learn language to teach computer programming. When structured languages came into vogue, Pascal, invented in 1971 by Niklaus Wirth, quickly became a favorite in computer science circles. Meanwhile, programmers handling complex jobs such as artificial intelligence favored LISP.

Less elegant and formalistic than Pascal, the language called C, developed at Bell Laboratories, took over many programming tasks during the 1980s. All of the programs constituting the UNIX operating system, the "housekeeper" used on many high-powered workstations, are written in C. A language called Java, created by Sun Microsystems, is the "hot" language of the Internet, as Web browser programs are given the capability to load programs written in Java off the Net, translate them into the machine language of their computer, and get them to work. The extension to Java is networking software from Sun that will allow disparate computers to share their processing power by connecting them readily over a network. This software would connect a wide range of devices—from cellular phones to thermostats.

## Creating Order out of Chaos: The Rise of Structured and Object-Oriented Languages

Although the earliest programmers had only primitive tools for creating the machine instructions that ultimately tell computers what to do, they were able to keep up with their tasks, in part, because the machines they programmed had much smaller memories. An accounting program in 1964 might take up only 1,000 bytes compared to the tens of millions of bytes for today's word processor or spreadsheet or database program.

Over the years, how software is written and organized has changed just as much as the computers on which the programs run. Big, powerful applications are possible not just because memories have ballooned and processors quickened. Also indispensable have been software writing techniques for dividing up chores and relying on code written by others. These enormous programming projects must be carved into manageable pieces that can be worked on independently, then brought back to perform in concert. For a long time, many big programs were written as a long list of sequential commands—one executed directly after another by the CPU. Some flexibility was given to those programs by invoking smaller programs called *subroutines*, which allowed a computer to jump from one part of a program to another.

If a task was complicated, however, programs would jump around from subroutine to subroutine, creating what became known as "spaghetti code"—a tangle of procedures impossible for programmers to understand and thereby revise. Large projects faced major hurdles. It was difficult to divide a programming job into parts that individuals could tackle independently. How one programmer executed some piece of code could have unintended effects on the rest of the program and even crash it.

The initial answer was a technique called structured programming, first popularized during the 1970s. Structured programming allowed big tasks to be divided up into smaller projects that could be safely worked on independently. The classic example of a structured program was the PASCAL programming language, although the C language also could be used to write structured programs. These techniques allowed big tasks to be divided up into smaller projects that could be safely worked on independently. Programmers relied on the notion of "information hiding"—each part of a large program performed a carefully defined task so programmers no longer needed to know about the details of other parts of the system: they only had to know what input data to supply and what output to expect. Information hiding allowed teams of programmers to work independently, and each subprogram could be used again and again. Even today, however, some programmers still do not use structured-programming techniques, and increasingly complex programs tax the ability of programmers to make all their parts work together.

## Computing in the '90s: The Evolution of the Icon

The average computer user who "points and clicks"—uses a mouse to click on a picture (or icon) that represents a particular function—is probably happily unaware of the revolution in programming that came into being during the 1980s. (By contrast, most users are very aware that their new generation of programs use enormous amounts of computer memory and storage space, having probably already hit the "disk full" wall when loading one too many computer games onto their hard disk drive.) In computer programming jargon, the icon is known as a *graphical user interface*, or GUI, its best-known incarnation being the software "desktop" on an Apple Macintosh computer or in Microsoft Windows. These graphical representations have also become a central component of the browser software used to link together pictures, text, audio, and video on the World Wide Web (see chapter 3).

Central to the use of GUIs is the high-resolution monitor screen, which is "bitmapped." Bitmapping assigns every area of the screen an "address" analogous to a location on a square of graph paper designated by a pair of coordinates. Each square is a pixel, or picture element. The number of pixels depends on the resolution of the screen, but each can have an x-and-y address that is described in bits. The computer can thus keep track of the position of any graphic and word, as well as the current location of the pointer being moved around by the mouse. This bitmap screen allowed the introduction of graphical objects with programmed characteristics, called icons. The functions embodied in these intuitive user interfaces lent themselves to

**Figure 2.2**   The graphical user interface employs icons to carry out commands such as opening files. This screen, created using Microsoft Windows 95, shows a series of windows with icons in the background.

structured programming and a newer set of methods, called *object-oriented programming*. This latter technique uses computer languages that help programmers break programs up into units that are entirely self-sufficient, with a common system for exchanging information between the units.

These self-contained units of software, called *objects*, are essentially complete small programs in themselves. They create the structure they need in memory to perform their required tasks, along with the functions needed to manipulate that data. And they are available to be used by any other object.

In a graphical interface created using object-oriented programming, an application can call up smaller objects that control important aspects of a metaphorical desktop. Each icon would have an associated object, so that a file folder might represent a directory of files and a piece of paper an individual file. When writing a new program or expanding one, a programmer can simply incorporate the functions of existing objects, such as displaying a file as a piece of paper that can be opened with a mouse click. Thus the programmer doesn't have to rewrite code. And because these objects all talk with each other in the same way (in theory), they should be able to exchange data easily. The objects all become components of one collective program running on your computer.

## Teaching Machines to Think:
## The Ups and Downs of Artificial Intelligence

For most people, the revolution in computer software design has resulted in enormously useful tools that can perform such relatively mundane functions as balancing a budget or running an educational game. Computer scientists, meanwhile, are exploring more esoteric realms that already play an important role in weather forecasting or predicting ups and downs in the financial markets. One area that has fascinated computer cognoscenti for decades is that of artificial intelligence, or AI. If we've become more adept at telling the computer what to do, can the computer, in turn, learn to do more? What if we could combine the awesome data-crunching abilities of the computer with the attributes of human thought processes?

As repositories and processors of raw information, computers have fulfilled most expectations. But the quest to make the computer "more human"—both in the way it interacts with the outside world and in its ability to think creatively on its own—has had a much more checkered history. In the 1950s, many computer scientists were confident that in only twenty years their machines would be "conscious" and conversing with us in natural language, thanks to AI. After forty years, both of these goals seem a long way off. On the one hand, computer advice is helpful in such endeavors as making medical diagnoses. On the other, to communicate with a computer, we still usually must tap a command on a keyboard or click a screen symbol with a mouse.

Computers that do dispense useful advice are called *expert systems*. Since the 1980s, expert system software has advised factory managers and helped doctors diagnose disease. The computer languages used to program expert systems are virtually all forms or descendants of one programming language called LISP, developed by AI pioneer John McCarthy in 1958. LISP works by allowing programmers to link endless lists of related information in a computer's memory. Users then can search for and retrieve patterns in these lists or call up desired pieces of them. Thus, a programmer can tell a computer that Socrates is a man, all men are mortal, and the computer will be able to use the rules it has been "fed" to also know Socrates is mortal.

The core of an expert system is its knowledge base. It differs from a database, which will have repositories of repetitive information such as address lists or invoice records, because it includes information that tells a computer what to do with these records or how these pieces of information are related to each other. A medical expert system might ask a doctor a series of questions about a patient's condition and then, using the rules in its knowledge base, recommend a course of treatment: The patient has a urinary tract infection. If it was acquired in the normal walk of life, it is probably an *Escherichia coli* infection that is treatable with orally administered antibiotics. If the patient acquired the infection in the hospital, it is more likely to be by a resistant bacterium, requiring intravenous antibiotics.

## Enter Heuristics

An expert system also strives to "teach" a computer to function in a world that comes in shades of gray. Usually, computers work through problems step-by-step by following a string of precise commands called an *algorithm*. Human beings are more apt to govern their choices and actions by rules that are less certain to provide a right answer. Part of building an expert system's knowledge base involves loading it with these less certain rules called heuristic rules. While a database is strictly a library of factual knowledge, an expert system comprises both factual and procedural knowledge—the latter being rules for how to use the factual knowledge. An expert system also has what is called an *inference engine*—a system for working through and drawing conclusions from its knowledge base.

Two practical problems have dogged the usefulness of expert systems. First, they take an enormous amount of labor to construct, which must be done largely by highly trained and skilled "knowledge engineers" who come up with heuristics by watching human experts at work. Second, they are tough to maintain and update, as they must be supervised by these same computer sophisticates. Other practical problems include a total lack of understanding about anything a particular expert system has not been instructed in, and an inability to convey what it knows to other expert systems; there is no knowledge sharing between them.

## HEURISTICS AT WORK:
## OR, HOW COMPUTERS "EVOLVE" TO BECOME
## TOP-NOTCH CHESS AND CHECKERS PLAYERS

The first experiments in computer game playing were conducted by IBM researcher Arthur Samuel in 1952. His early checkers-playing programs were so successful that pioneers in AI—a field that had just found itself a name—predicted there would be a nonhuman chess champion sometime in the 1960s. They've had to wait longer than they expected, but dedicated chess computers such as IBM's Deep Blue are at last up to the task. In 1997, Deep Blue beat champion Garry Kasparov in a six-game match in which the computer won two games and drew three others.

Arthur Samuel's checkers program employed most of the techniques game programmers still depend on, including the use of heuristic rules to develop a playing strategy. Samuel realized that even checkers presented too many possible combinations of moves for a computer to work out each possible outcome. Indeed, he estimated that even if his computer could look at three choices every millisecond, it would still take it more than one hundred years to work out all possible combinations. He needed to come up with rules that would drastically limit the number of moves considered. Because it would be governed by rules not certain to bring success, looking for the next move would be a heuristic search rather than an algorithmic one.

Samuel let the computer sharpen its rules for evaluating the strengths of moves and positions by having two versions of the program play each other. One was called the alpha version, which altered its heuristic evaluation formulas before every match. This went up against a beta version, the version that had won most recently without changes. When the alpha version finally came up with modified heuristics that allowed it to beat the beta, the beta version then had its logic updated and the process was repeated. Hence, because the program learned on its own, it is said its heuristics "evolved" rather than were designed.

Samuel's program sometimes needed "patches" of code thought out by a human when it got stuck trying to beat certain techniques and strategies, but after four years it proved itself a top-notch player. Similarly, chess programs have needed human intervention to help them play more creatively and unpredictably. Yet the core of these programs, their heuristic formulas, have typically "evolved" into equations too complicated for human programmers to understand.

## Talking Machines: The Quest for Natural Speech and Common Sense

In 1950 the brilliant digital visionary and mathematician Alan Turing outlined a test for evaluating whether a computer really thinks. Place a person in a room with two terminals. One terminal leads to another person with a terminal. The second terminal leads to a computer. Let the person ask any question he wants over the terminals for five minutes. If he can't figure out which terminal leads to the computer, that computer can be said to be thinking. Turing predicted that by the end of the century, computers would be so fluent in language and knowledgeable that they would be able to pass this test. No one then could seriously question their ability to think.

Unfortunately for this prediction, human language has proven to be exactly the kind of real-world arena in which computers are intractably gauche. Big hurdles in language comprehension include the human penchant for idioms—expressions that mean things different from the sum of their words—and words with numerous meanings. In this way, computers are rather like Amelia Bedelia, the fictional maid who looked for a furry bear every time she was instructed to "bear left" while driving a car. Attempts to make machines either understand everyday human expressions or translate from one language to another have often foundered. In the 1960s the Pentagon abandoned all efforts to come up with an automatic Russian/English translator after the task proved overwhelmingly difficult. In the 1970s it sought to develop a computer that "spoke" natural English, but again found this was too hard a project. The Japanese government in turn launched what was called its Fifth Generation initiative in the 1980s. This was designed to develop a computer that could understand natural language, do translation, and reason like a human. That project also failed. The natural-language capabilities of "adventure" computer games, in which the computer responds to simple, declarative sentences, and awkward grammar checkers remain state-of-the-art "language processors."

Computers can be adept at mimicking human speech patterns, but this does not mean they are responding to human language in a significant way. In 1966, Joseph Weizenbaum developed the shrink-in-a-box program ELIZA, which featured clever responses to human questions so that the computer sounded like your local Manhattan psychiatrist. The trick was to have ELIZA spit out so many platitudes and vapid phrases that they made sense even when she hadn't a clue to what you were talking about. The program even knew to ask for payment at the end of a session.

Computers have made progress in some areas. Automatic telephone operators ask for and understand spoken phone numbers. Speech-recognition programs can recognize and respond to caller inquiries about airline schedules for more than 1,000 cities. Software can translate a speaker's words into text with reasonable accuracy, but only after the software has received intensive exposure to the voice of a particular individual. In addition to mouse and keyboard, limited control of a computer through spoken commands is already possible. Computer scientists even now predict they'll achieve the ability to automate translation from one language to another in the next ten years. It is still hard to know, however, when a computer will be able to respond to the meaning of a speaker's words in the way the infamous HAL did in Arthur C. Clarke's *2001*.

There is now an annual competition for the computer program that can come closest to passing the Turing Test. Recent winners include a sex therapist program and one that makes a computer sound like a politician. (Judges have also misidentified human participants as machines.) These programs "understand" natural language to the extent that they can make what we humans perceive as reasonable responses to natural language. But they are not considered truly intelligent because they cannot discuss anything outside their limited area of expertise.

Some computer scientists are trying to fill in this deficit. In Austin, Texas, "knowledge engineers" are loading a computer with every rule they can come up with that may be deduced from human experience, hoping that together these rules will give the computer common sense through having a knowledge base on par with a human, which will allow it to understand normal language. The idea is that what has so far separated computers from attaining common sense and ability to understand humans is the shared information that we bring to any sentence we read or utter.

The scientists of what is called the CYC project—short for enCYClopedia—began by trying to give the computer the set of rules that would allow it to understand all the nuances attached to the sentences "Napoleon died on St. Helena. Wellington was sad." It took them three months just to input the rules these events implied—they told the computer who Napoleon was, what a human being was and what it means that we die, and on and on. Begun in the mid-1980s and originally scheduled to take ten years, the CYC project is now expected to be finished sometime early in the next century.

Some AI researchers have rejected this catalog-like approach. Instead of loading up a computer with facts and rules, Rodney A. Brooks at the Massachusetts Institute of Technology's Artificial Intelligence Laboratory has been building a self-teaching robot called Cog. The robot—which consists of a head, torso, arms, and hands—is supposed to develop innate intelligence by teaching itself about the world through interacting with its surroundings. The concepts of "up" or "down" evolve from processing inputs from the robot's sensors. The idea for Cog came about when Brooks began to contemplate what it would take to create a computer with the capabilities of HAL, the legendary artificial brain of Arthur C. Clarke's *2001: A Space Odyssey*.

Before Cog, Brooks had built insect robots that taught themselves to walk, not through any canned set of rules, but by triggering a sensor to provide "negative feedback." Every time the six-legged artificial cockroach fell down, it received a kind of shock. Repeated shocks trained a neural network program until the roach "discovered" how to keep three legs down while it moved three others. The software first learned simple actions, lifting a leg, and then added more complex motions and behaviors—such as avoidance of other objects.

Teaching machines common sense is also key to a new kind of program that some AI experts are predicting will revolutionize the way we work with the computer in the coming decades. The idea is to reduce the tedious amount of time many people must spend at a computer doing essentially menial tasks, e.g., searching databases and on-line services for

**Figure 2.3** Cog, a self-teaching robot in development at the Artificial Intelligence Laboratory at the Massachusetts Institute of Technology. Cog is intended to develop intelligence through interaction with researchers in the same way that a child learns from the people in his or her environment.

information, and generating and writing routine reports. Instead of invoking the programs needed to do this work, a computer will call upon an agent—a digital program with enough smarts and common sense to do all the grunt work for you.

Agents will be, it is hoped, the concierges of the online services and personal electronic butlers. They will study your habits when using a computer in order to learn about your interests and needs so they can repeat these actions. For example, one will know you like reading about gardening, but only about plants you can grow in a window box. So it will search far and wide for gardening articles—far more widely than you'd possibly have time for on your own—and cull down its findings to only the stories about plants that could be grown in window boxes. Some agent programs have already begun to roam the Internet, gathering information customized to an individual's personal tastes.

---

### COMPUTER SIGHT: A LONG WAY TO GO

In the 1980s the Pentagon spent over $100 million trying to develop what it called an Autonomous Land Vehicle or ALV, a self-guiding "delivery truck" that could peer out over oncoming terrain, make sense of what it saw, and respond accordingly. Such a machine would save humans from having to be in the driver's seat during risky wartime missions such as delivering ammunition or doing reconnaissance. ALVs were designed to handle open desert. The machine was to have enough intelligence and responsiveness to traverse this terrain at speeds faster than a running soldier.

Unfortunately, the amount of visual information to be interpreted proved overwhelming for even a vanful of computer equipment. After six years the AVL project was killed when the prototype could cover only six hundred yards of open terrain while traveling at three miles per hour. Autonomous vehicles have done better on the open road than in the field. In tests, computer-driven vehicles have logged hundreds of highway miles without needing intervention by their human drivers.

---

## Understanding an Imprecise World:
## Neural Networks and Fuzzy Logic

If computers always think in terms of yes or no, on or off, ones or zeros, is it really possible for them to come to grips with a world full of ambiguity and nuance? Can a machine as literal-minded as the computer master subtlety? In the past two decades, two techniques have assisted computers in coming to grips with our gray, poorly defined world.

One technique is to organize a computer's memory into elements designed to imitate the behavior of living neurons in a brain. Called *neural networks*, these information processors learn to recognize underlying patterns in visual, numeric, or other data by generalizing from examples. For instance, they can pick out the account number on a specific customer's personal check, but they do this only after being "trained" to recognize the number after exposure to many different checks with slightly varying configurations of numbers. This ability to generalize means that such systems can work with "messy" data: a check number that has been scribbled over with a pen, for instance. Neural networks are proving useful in tasks as varied as recognizing handwritten letters and identifying precancerous cells in a Pap smear. They are also proving helpful in monitoring chemical processes and in economic forecasting.

A neural network "learns" how to recognize a face, handwriting, or a check number through training. The units that mimic nerve cells are arranged in layers of processing units. All the "stimulated" units in the bottom or input layer pass on digital values to an internal layer of units to which they are logically connected; these then "decide" whether they should continue to send the signal along, depending on whether their stimulation reaches a given threshold. Through this filtering, the output units in the final layer develop a general picture of the original image—the network learns, for example, to recognize that a check number has a certain length, though it may vary slightly from time to time. It also can detect number spacing and where the numerals are located. If the final image is deemed "correct," the values exchanged between the cells that interacted to help create it will be strengthened by having their values increased, while incorrect connections are reduced or weakened. After encountering and having ties between cells rewarded or reduced enough times with different examples, neural networks are able to identify a symbol from various examples full of variations. They can be taught to recognize a handwritten *e*, for example, from the numerous versions that might be scrawled by different writers, with no two exactly alike.

The real importance of neural networks is that they can learn to perceive patterns from disparate events that can give them predictive and diagnostic power. Feed them events and link them to some consequence, and over time the network will be able to predict consequences occurring when new events, filtered through a network, lead to the same conclusion. With this technique, neural networks have been applied in fields as diverse as stock market predictions and disease diagnosis.

## HOW A MACHINE PROVED
## A THEOREM THAT PERPLEXED HUMANS

In 1996 a computer showed it could come up with a solution to a mathematical problem that had confounded human mathematicians for fifty years. The problem, a pure mathematics exercise involving Boolean algebra, was explored by a computer that tried various solutions for eight days before it flashed to its programmer, mathematician William McCune, the message, "HEY McCune, WE HAVE A PROOF."

Teaching a computer how to prove hard math theorems is similar to teaching one how to play chess. First steps in this area were taken in the 1970s, when researchers began to experiment with computers to see if they could solve mathematical problems humans had already conquered. As with a chess move, the number of approaches to a problem was too vast for them all to be explored systematically. Heuristic strategies had to be taught, such as rule-of-thumb techniques that had successfully solved other problems. Working with problems that had known solutions allowed mathematicians to see where the computer got bogged down, and adjust its strategies accordingly. Eventually the computer was on its own, tackling problems that had eluded a human solution. Human advice, combined with a computer's unflagging ability to keep on trying again, created a formidable player, this one on the mathematics scene.

Another technique, "fuzzy logic," is a way to program computers to respond appropriately to the fact that the world is rarely unambiguous. This is essentially an averaging technique that responds to the degrees of a condition, and frees computers from the bind of looking at the world as a collection of either/or conditions. When computers only respond to conditions as either existing or not existing, they lack any subtlety. A better approach is to tie a response to the degree to which a condition exists.

An example of fuzzy logic in action would be the fine-tuning of a small device that is designed to maintain the chlorine level of a swimming pool. A device controlled with traditional logic will sample the pool water periodically, and do nothing until the chlorine drops below some threshold level, when it will "decide" it is "dechlorinated" and dump in a prescribed amount of chorine. A fuzzy-logic device will measure the water, decide to what degree the pool is "dechlorinated," and respond by adding an amount of chemical proportional to this level. If the pool has a chlorine level halfway between what is considered "fully chlorinated" and "not chlorinated," it will add half the chlorine it would if it were "not chlorinated." If responses involve different processes—machines running on "high" or "medium" or "low"—the fuzzy action will be more complicated, but it will still boil down to deciding on a response proportionate to the degree of some measured condition or conditions.

Fuzzy logic has sharpened the performance of various monitoring systems around the world. Fuzzy systems not only perform well in controlling household items such as air conditioners and refrigerators, but can greatly increase the efficiencies of mechanical systems as diverse as cement kilns, vacuum cleaners, camcorders, and washing machines. Fuzzy logic is seen as so effective and as having so many applications that dedicated fuzzy chips designed to work with fuzzy sets and rules have been developed. In Japan, fuzzy-logic controllers run subway systems more smoothly than human drivers, who now take over only during off-peak hours to keep in practice.

## The Splendid Simulator: The Promise of Virtual Reality

Though humans have had only mixed results in getting computers to comprehend the realities of the outside world, they have had great success in reproducing the outside world on their computers. Computers may have trouble responding to the complex, logically confounding stimuli of the real world. But they can project images of the real world brilliantly and adjust these images in response to controlled stimuli. In essence, they can mimic real life, even if they can't really cope with its ambiguities.

The creation of simulated worlds—often called virtual reality—may be the arena in which computers have come closest to fulfilling the fantastic roles science fiction predicted they would have in our lives. Before the 1991 Gulf War, the Pentagon constructed a "virtual Kuwait" to train on and test equipment. Medical students now have "virtual cadavers" to supplement the "live" ones of their anatomy classes. New car models are routinely taken on a virtual "test drive" long before any of their parts are tooled. In a complicated program such as the Boeing 777 development project, hardly any of this aircraft's millions of pieces weren't run through the simulation wringer before the first prototype was constructed.

In some cases, these simulations combine software capable of simulating three-dimensional images with specialized hardware that allows the user to enter a realistic environment. Motion sensors wired to a helmet detect the user's movement and alter scenery and sound on stereoscopic eyepieces and headphones. Sensor-laden gloves "touch" an object by providing a sensation of force as a hand moves. Thus a visitor to a

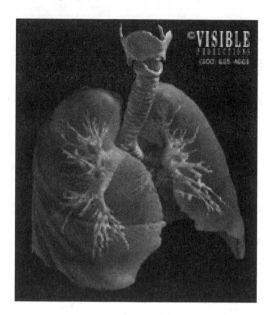

**Figure 2.4** A virtual reality re-creation of the human lung allows students to study anatomy without resorting to scalpels and cadavers.

baseball simulator might pick up and pitch a virtual baseball, watch it sail across home plate, and listen to the loud smack in the catcher's glove.

Simulations have broad-reaching uses. Because they are so willing to repeat tasks endlessly, no matter how complicated and tedious, computers helped to invent the science of chaos theory: their tireless calculations demonstrated how tiny changes in variables in complicated systems such as the world's weather may have enormous, unintended effects, making these systems fundamentally unpredictable. Neural networks are not only allowing computers to filter this fuzzy world into the black-and-white images of their liking, but are also providing electronic models of how biological neural networks and brains work. Biologists are modeling ecosystems, while climatologists use enormous computer models to allow them to look at the possible consequences of global warming, not to mention nuclear war.

## Artificial Life

One field that has proven fruitful for both biologists and computer programmers is the study of artificial life. Central to many AI applications,

**Figure 2.5** A helmet loaded with sensors that can detect movement and alter scenery and sound on stereoscopic visual and hearing systems presents the subject with a virtual world. Using the sensor-laden data glove, the subject manipulates, operates, and controls her experience.

from getting a computer to improve its chess game on its own to modifying its methods for trying to solve a math problem, is the idea of the genetic algorithm, a technique first tried out in Arthur Samuel's checkers program (see page 42). Algorithms and heuristics are programmed to alter themselves slightly or mutate. The effectiveness of new and old is compared. The version that performs best will then engender new versions that get tested against it, and so on. Software improves itself by being subjected to the pressures of Darwinian survival of the fittest. In essence, this method of software development involves randomly trying everything and then allowing what works best to stick around. Sheer luck can come up with algorithms and techniques humans will never think of.

Artificial life on computers also serves as models for how self-replicating ordered systems can evolve and change over time. Do populations remain stable and then undergo sudden spurts of evolution, in what paleontologist Stephen Jay Gould calls "punctuated equilibrium"? Or do forms adapt slowly and evolve gradually, according to the more traditional view? Simulated life can go through thousands of generations in a blink of an eye to help answer questions like these. You can see the history of the world in not much more than a grain of silicon.

# MIND BENDERS:
## A GLOSSARY OF COMPUTER SOFTWARE TERMS

**algorithm**   A clear, unambiguous system of steps that, if followed, are guaranteed to solve a problem.

**artificial intelligence (AI)**   A field of computer science interested in endowing computers with the capacities of human thought.

**assembly language**   A computer language that ties every machine code instruction with a wordlike mnemonic, which, before running, is translated into usable machine code.

**compiler**   A special program that takes instructions written in a higher, more humanlike computer language and generates a separate, runnable program in machine code.

**expert system**   A computer knowledge base that holds not only facts about a given field but also rules for dealing with these facts.

**fuzzy logic**   An averaging technique for looking at the degree of various conditions so that an appropriate response may be taken.

**graphical user interface (GUI)**   A method of easing interactions between computer and computer user by employing visual cues and icons instead of teletype-like words and questions to gain input from the user, usually requiring a "mouse" or similar input device for clicking selections.

**heuristic rule**   A rule of thumb given a computer that is generally but not necessarily always true.

**interpreter**   A special computer program that converts instructions written in a higher, more humanlike computer language into machine code that is sent immediately to the CPU.

**machine code**   The system of ones and zeros that the CPU interprets as sensible instructions.

**neural network**   A method for identifying trends and patterns by filtering information through a series of electronic circuits organized like biological nerve cells.

**object-oriented programming**   A software development approach whereby all pieces of a program are self-contained and communicate only through strictly defined channels. Each piece may also then be used by other programs.

**operating system**   A collection of programs that tells a computer how to run everything outside the CPU, including how to organize memory. This includes the necessary housekeeping software that runs the computer and that other programs will invoke to perform simple tasks.

**structured programming**   A technique for allowing large programs to be broken up into self-contained pieces that can be designed independently from the other pieces.

# 3

# Wiring the World: Telecommunications and Data Networks

The personal computer of the 1980s replaced the typewriter, the accountant's ledger books, and the closely guarded secrets of the corporate mainframe. The real upheaval, however, arrived a decade later, when the information contained in each isolated machine could be piped routinely into the wall and transmitted anywhere. Thus an electronic mail message could be relayed to the order counter at the corner pizza shop, or elementary-school children could lob questions to space shuttle astronauts. These networks represent more than just the novel spectacle of queries to astronauts about the constituents of a zero-gravity breakfast. Many of the functions of communications technologies, from telephone to mail services, have begun to merge. Electronics makers now have the means to craft a single hand-held device that can relay phone calls, accept electronic mail, provide paging services, and—when its owner has gone astray—make a quick call to a satellite positioning service that will convey travel instructions.

In its most basic form, a network is simply a set of connections that may be linked together by a variety of physical media, whether telephone lines, satellites, radio connections, optical fibers, or the coaxial cables used to transmit television programs. Data communications have assumed increasing importance as the world has come to recognize the possibilities offered by linking one computer to another. In a period of less than five years, the Internet progressed from an arcane haven for academics and military types to the global medium for connecting one computer with another. As the capacity of the Internet has expanded, this information conduit has begun to subsume many of the functions formerly fulfilled by traditional media. Today television networks, newspapers, and magazines are obliged to supplement their standard offerings with World Wide Web pages. In coming years, the Web may evolve into the main vehicle for their delivery of content. The Internet draws some of its attraction from interactivity: an ability to let people talk back immediately to a newspaper or, if so inclined, to start their own vanity newsletter or a Web page that shows their stamp collection to the world. And electronic mail holds the promise of reviving—in a somewhat supercharged fashion—the ancient art of letter writing. As much as they are a vehicle for entertainment, information, and communications, the Internet and online communications draw their allure from their potential as a direct sales outlet to consumers, hawking everything from stocks to socks.

The transformation of the old-line media establishment progresses in lockstep with the emergence of compact and mobile computing technologies whose connection to the world may be a handheld telephone with a wireless radio link to a satellite that allows voice or data calls in either Los Angeles or Mogadishu. While achieving global reach, the technology may literally begin to move closer to home. Over time, "smart homes," controlled by internal data networks, may automate a sprinkler system or be fitted with motion sensors that turn on the lights and play a favorite CD when someone enters the room. The notion of exchanging data anywhere, at any time, may even extend to one's person. Numerous schemes to fit one's eyeglasses, clothing, or even undergarments with networked computing devices have been put forward in both graduate student projects and even commercial products. Like the household robotic butler, some of these applications border on the fanciful. A BodyNet might sense sweat in an undergarment and then send a signal to turn up the air conditioner, though how big a market exists for wired underpants remains to be seen.

That imagination sometimes outstrips pragmatism can also be seen in the push to build an "information superhighway," a term that has become outdated as expectations for high-speed digital networks that transfer video and sophisticated graphics have arrived somewhat more slowly than expected. A wired world has also provoked international debate because of legal, social, and business dilemmas that may take years or perhaps decades to resolve. Academic and trade conferences abound in which social pundits ponder a host of questions: What does the notion of copyright and intellectual property mean when any digital object can be copied and relayed to the millions who frequent the Internet? How does one charge for digital content that has been traditionally provided over the Net without charge? And what means exist to protect the privacy of an E-mail message or the security of a credit-card transaction, or to shield one's children from a surfeit of Net-based pornography? What complicates these issues is that the prevailing ethos of the Internet culture borders on the libertarian, and national laws are difficult to enforce on a truly global network.

Still, the rise of computer networking has emerged as a watershed technological event as we cross the millennial divide. From a technical perspective, the most momentous developments in communications center on the gradual transition from analog to digital communications—and the rise of broadband networks capable of relaying enormous numbers of these bits. In this scenario, a stream of digital bits will intermingle transmissions of voice, video, and data, even to households and small businesses.

For the moment, however, the vast majority of homes in the United States still receive their phone calls as analog signals, in which the waveform of the electrical signal varies over time in the same way that a sound wave changes as it travels through the air. In an ordinary phone call, analog signals' variations in amplitude and frequency correspond to the loudness and tone of the human voice. When a signal reaches its destination, it makes the telephone's speaker vibrate, re-creating spoken sounds. Most radio and television broadcasts also travel as continuous waves of analog information. To send or receive data from a computer over analog lines, the computer must be equipped with a modem—a modulator/demodulator—that converts a series of zeros and ones to an analog signal or transforms it back into a stream of bits.

This analog technology, with roots in the nineteenth century, is gradually being supplanted by digital communications systems in which the qualities of voice or picture image are represented as binary information—bits representing values of zero or one (see chapter 1). Computers can manipulate those bits in order to move information from place to place or interpret the data to re-create an original sound or image. International communications are still a hybrid of digital and analog systems. Most traffic on the trunk lines between phone company switching offices travels as digital pulses. But the "last mile" connection to your home is still likely to be an analog transmission.

Going digital has considerable benefits for the entire spectrum of communications. Analog signals suffer more easily from corruption that results from interference with other signals—and digital integrated circuits used to process the signals are cheaper than their analog counterparts. Digital communications are also clearer than analog because of the error detection and correction capacities of the software and hardware. Another important feature of a digital network is the ability to intersperse varying types of information, including text, pictures, sound, and numerical data on the same line. Network capacity also increases dramatically, since digital signals can be compressed, allowing them to carry more information. Converting satellite communications into digital form has increased capacity by an estimated factor of five.

The capacity of a digital network is often described in terms of the number of bits that the network can transmit from place to place every second. A digital telephone network carries millions of channels of voice conversations, typically transmitted at 64,000 bits per second. This is enough capacity for a telephone conversation, but far too little for VCR-quality television, which needs about 1.5 million bits per second. High-definition television requires even more: some 20 million bits per second. The capacity of a digital network, measured as the number of bits it can transmit every second, is called *bandwidth*. Engineers refer to networks with the capacity to send only a relatively small number of bits as low-bandwidth or narrowband networks. High-capacity networks that can transmit a torrent of bits are called broadband or high-bandwidth networks.

The components of the networks continue to undergo dramatic evolution so that even the copper wires that have carried phone conversations since the era of Alexander Graham Bell may one day find themselves obsolete, as fiberoptic lines or wireless links give rise to broadband networks. Until then, methods of taking advantage of digital communications using the existing base of copper wiring have been devised. An Integrated Services Digital Network (ISDN) allows voice, data, and some types of video to be transmitted over the same phone line at speeds of 128,000 bits per second, more than twice as fast as commonly used high-speed modems. Some telecommunications analysts even question whether it is necessary to go to the expense of creating expensive new infrastructure—putting in place high-speed fiber. The reason: another digital standard—under the rubric of Digital Subscriber Line (DSL)—offers enough speed, up to a million bits per second or more on regular copper wire. A number of U.S. computer companies and international telecommunications carriers have embraced the technology. Besides DSL, a broadband network already extends into a majority of American homes. Specialized modems that can send Internet data over the television cable network at a speed of millions of bits per second have already been adopted in tens of thousands of homes—and the technology, in theory, might suffice for the delivery of high-speed interactive data and video.

## THE ELECTROMAGNETIC SPECTRUM

Communication signals, whether bouncing off a satellite in the ether of space or coursing through old-fashioned telephone wire, travel from one place to another utilizing portions of the electromagnetic spectrum. The radiation that makes up the electromagnetic spectrum travels at about 186,000 miles per second in a vacuum, producing oscillating electric and magnetic fields as it goes along. Light, and other forms of energy, including radio, microwaves, and ultraviolet radiation, are components of the electromagnetic spectrum. We usually categorize electromagnetic radiation according to its wavelength and frequency. The wavelength is the length of a complete up-and-down modulation of the wave—the distance between two wave crests. The frequency is the number of these complete cycles that will pass a point within a specified unit of time. Since all electromagnetic radiation travels at the same speed, the shorter the wavelength, the higher the frequency. Visible light makes up a tiny portion of the full range of electromagnetic radiation.

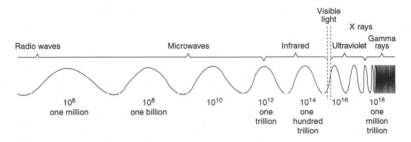

**Figure 3.1** The electromagnetic spectrum encompasses the full range of radiation. The spectrum ranges from low frequency signals like radio waves to highly energetic gamma rays. Visible light occupies a small fraction of the spectrum.

The frequency of a wave is usually expressed in hertz. In the case of a radio wave one mile (1.6 kilometers) long, about 186,000 complete waves will pass a point in a given second, so the frequency of this wave is 186 kilohertz. (One kilohertz—or kHz—is a thousand modulations per second. A megahertz—or mHz—is a million.) Visible light typically has a wavelength of 400 to 700 nanometers, and its frequency is roughly 400 to 750 terahertz per second (a terahertz is a trillion hertz). From the shortest to the longest, the spectrum of waves includes gamma rays, X rays, ultraviolet, visible light, infrared (heat) waves, microwaves, and high-frequency signals used for television and ordinary radio as well as beamed short-wave radio communication and the longer-wave frequencies used to transmit voice signals. Optical fiber communications take place in the near infrared portion of the spectrum, with wavelengths of .8 to 1.6 microns (a micron being a millionth of a meter). The electromagnetic waves used to broadcast radio and television have wavelengths ranging from hundreds of meters down to fractions of a centimeter. Microwaves, the shortest of the radio waves, are used to transmit radar and satellite signals.

Use of the electromagnetic spectrum is regulated by government agencies around the world. In the case of the United States, the governing body is the Federal Communications Commission, or FCC. The FCC allocates or auctions off portions of the radio frequency spectrum to communications companies that make use of or sell communications services. The competition for pieces of the spectrum can be fierce, and as modern communications increasingly take to the airwaves, there are concerns that commercial uses will crowd out the needs of scientists and radio astronomers (see page 66).

## Communication Pipes

The metaphor often used for building a web of global communications is "wiring the world." And indeed most communications still take place on a pair of copper wires. But, ironically, one of the key symbols for this new era is not a traditional wire but a thin fiber made of ultrapure glass. This optical fiber, which carries information as a modulated beam of light, is a key technology in implementing broadband communications. A broadband network will help transfer digital audio, graphics, and video and computer data rapidly enough that the user will not notice any waiting after requesting information from another computer.

Though most households make telephone calls and Internet connections over copper wiring, most telecommunications companies have already replaced many of the connections on the trunk lines between their central offices with optical fibers. Optical fibers can also be used in the high-speed local area networks that transfer information within a single office complex.

The cost of laying fiber to individual homes—in the absence of critical new services that consumers are willing to pay for—has, for the moment, stalled plans to make fiber the medium of choice for ubiquitous broadband communications. In fact, some communications experts believe that broadband links to the home may be completed with a wireless connection. Alternatively, the cable television network could provide the needed high-speed connections.

Whatever scenario comes to pass, fiberoptics will still play a role when high-capacity networking is required. Fiberoptics can transmit information at rates and volumes that far outstrip conventional copper wiring. Hair-thin glass fibers that transmit light from lasers or light-emitting diodes offer the possibility of virtually unlimited communication capacity at relatively low cost. The amount of information that can be carried depends on the frequency of the signal: the higher the frequency, the more bits fit in what are sometimes called "light pipes" (see page 57). A single fiber has enough capacity to accommodate millions of telephone calls or thousands of television channels, although only a fraction of the bandwidth is utilized in today's networks.

By contrast, most copper wiring still sends data at rates that do not usually exceed 50,000 bits or so per second. Fiber's capacity steadily increases as the technology improves. Following the same trajectory as microprocessor technology, the capacity of optical fiber to transmit information has often doubled every year since the 1970s, when the first major fiberoptic line was installed. This doubling is accomplished by improvements in the electronics, transmitters, and receivers.

The advent of fiberoptic communications was preceded by decades of laborious and hugely expensive research. Engineers confronted the

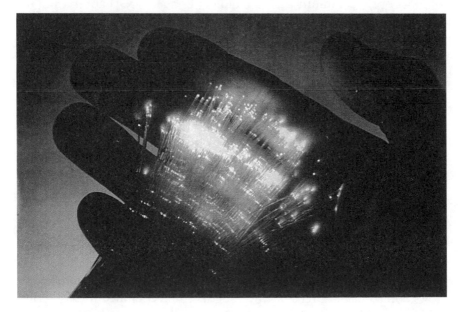

**Figure 3.2** Fiberoptic cable consists of hair-thin glass filaments that transmit information using light waves.

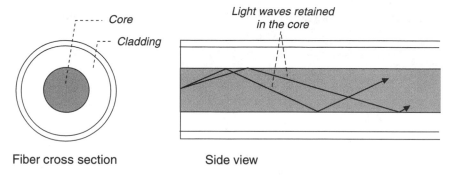

**Figure 3.3** An optical fiber consists of a core and a cladding. The slightly different refractive index of the cladding keeps light waves inside the core region. The refractive index of a material is a measure of its ability to bend light.

challenge of developing a glass that did not "lose" the light waves it was transmitting. The solution was to develop a fiber that had an inner core and an outer surface, or *cladding*, each of which has a different refractive index (the measure of its ability to bend light). When light moving through the core hits the cladding at a certain angle, it reflects back into the core; in essence, the cladding guides the light wave along the inner core. Fiber-optic transmissions in long-distance communications use a different scheme, in which light travels parallel to the cladding.

Fiber-optic communications require a light source that serves as the information transmitter, and receivers that can translate the messages back to an electrical signal at the other end. In fiber-optic transmissions, a device called a modulator acts as a shutter that opens and closes the transmitting light source—either a laser or a light-emitting diode—at rates that may exceed a billion times a second. Because of the increasing miniaturization of semiconductors, the transmitters and receivers used in fiber optics are so small they could fit together into a matchbox, while the light sources are no bigger than grains of salt. A commercial fiber system marketed in the late 1990s by Lucent Technologies had the capacity to fit 400 gigabits per second—that is, 400 billion bits—in a single fiber, enough to fit the entire worldwide traffic of the Internet. In addition, fibers capable of transmitting a terabit (a trillion bits) or more per second have been demonstrated.

The ability to transmit information at these astounding rates is a tribute to advances in technologies for transmitting different wavelengths of light on the same fiber. This technique, called wavelength division multiplexing, allows dozens or hundreds of wavelengths to be crammed into the same fiber, each relaying several gigabits or more every second. Wavelength-division multiplexing is a critical linchpin in all-optical networking in which signal processing occurs by directly manipulating the light itself. In addition to the multiplexers, optical amplifiers deployed in a telecommunications network strengthen a signal that has weakened after traveling tens of miles without optical-to-electronic conversion. One of the advantages of these devices, known as *erbium-doped fiber amplifiers*, is that they boost ultra-high-speed signals—those traveling at even tens of gigabits per second. In addition, simple optical switches that will guide a wavelength from one fiber to another have begun to be deployed by some telecommunications carriers.

## Taking to the Airwaves: The Rise of Mobile Communications

Fiber systems have their limitations in providing mobile service or supplying broadcast services over a large region or to remote areas. Rewiring the world with fiber is a costly undertaking. Tens of thousands of miles of cable, complete with transmitters, multiplexers, amplifiers, and photodetectors, will have to be physically installed underground and inside buildings. Another approach is to bypass physical wiring altogether and take to the airwaves. Wireless communications technologies—based on radio and microwave signals—cannot as yet match fiberoptics for transmitting billions of bits of information at lightning speeds. But wireless has many advantages. Mobile communications afford convenience, permitting "soccer moms" and international executives to place and receive calls while on the move. For developing countries that do not already have a communications infrastructure in place, there are significant cost and time advantages to using radio and satellite communications. Already, satellite operators are beaming digital broadcasts of movies, sports, and other entertainment into homes around the world at a fraction of the cost of building new cable and wiring infrastructure. And, to provide telephone service to the billions of people around the world who have never made a telephone call, developing countries are increasingly opting for wireless service.

## Space Calls: The Lure of Satellite Communications

The place where calls are relayed between a caller and its recipient is often found from hundreds to thousands of miles above the Earth, in the form of communications satellites. The hundreds of orbiting satellites are a linchpin of international communications: On an international telephone call, that slight time lag at the other end of the line results from the travel time of the signal to a satellite 22,300 miles away and back to the Earth.

Until the mid-1950s, the possibility of communication by satellite was considered to be mostly in the realm of science fiction. The intellectual father of the concept was Arthur C. Clarke, who, in 1945, wrote an article proposing the use of satellites for communications. In the decades since, communications satellites have experienced dramatic increases in capacity. Telstar 1, a communications satellite launched in 1962, carried only a few voice circuits. A contemporary satellite can relay many tens of thousands of telephone conversations or over 200 digital TV channels. Today, some of the advanced international telcom satellites launched by Intelsat, a consortium of the United States and more than one hundred other member nations, carry up to 100,000 phone channels.

Using a satellite for a telecommunication link requires only one repeater—the satellite itself—to amplify the signal. Some types of satellites can provide worldwide coverage with as few as three "birds," as satellites are sometimes called, parked 22,300 miles above the equator. By contrast, submarine or long-distance land cables need many repeaters. Putting a satellite to work in orbit above the earth is a far less labor- and materials-intensive task than laying tens of thousands of miles of submarine or land cables.

Most satellite systems communicate using microwaves that operate at frequencies ranging between 3 and 30 gigahertz (a gigahertz is a billion hertz). This band is not affected by electromagnetic interference in the ionosphere and does not get absorbed in the atmosphere. (Because microwaves do not bend to follow the curvature of the Earth, radio towers that transmit these signals for terrestrial communications must be spaced every 20 to 50 miles.) The projected expenditures for several current satellite projects for telephone and broadband data and video communications during the late 1990s will run into the tens of billions of dollars. There are also considerable risks associated with getting a payload into space: rocket-borne satellites must be placed in orbits that range from 400 to 22,300 miles above the Earth. Aborted or failed rocket launches have resulted in the destruction of communications satellites worth millions of dollars. A failure of a satellite can have dire consequences. When a communications satellite went down in 1998, the majority of U.S. paging services were lost for several days. But satellites generally perform as expected. After a launch, they are typically more than 99.9 percent reliable.

Once they have been placed in orbit, satellites use solar cells to power their functions and store the solar energy when the Earth eclipses the satellite's view of the sun. They are not equipped with powerful engines; rather, they stay in orbit because of the interplay between the Earth's gravity and their trajectory. A satellite in orbit is actually falling toward the Earth all the time because of gravity. But because the satellite is traveling very fast and at a high altitude, the Earth's surface curves away from

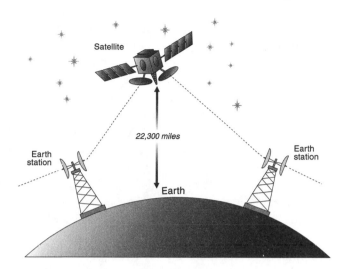

**Figure 3.4** Orbiting 22,300 miles above the Earth, this satellite transmits signals from one Earth station to another. An uplink sends data from the Earth station to the satellite, while a downlink from the satellite transmits the signal to the next Earth station. Although it is traveling at high velocity, this satellite appears to be stationary. Because of the satellite's distance from Earth, there is a half second round-trip delay between two Earth stations, which is noticeable in voice communications.

it as it falls. Thus, the satellite never actually gets closer. At times, on-board jets must be fired to ensure that the satellite's antenna and the solar array remain pointed in the right directions.

Communications satellites are classified according to their height of orbit. Many of the Earth's communications satellites that relay television and telephone messages orbit above the equator at a height of 22,300 miles. When these satellites are launched, they are accelerated to the right velocity and height so that their speed exactly matches the rotation of the Earth. Spacecraft that orbit the Earth at the same speed as the planet's rotation are called *geosynchronous* or *geostationary* satellites because they appear to be "parked" above a particular spot on the Earth's surface.

The advantage of such geostationary or geosynchronous satellites is that a telecommunications service provider can use a fixed signal to send signals to and receive them from the satellite, rather than having to follow a moving target across the sky. An equatorial orbit also has the advantage of covering an enormous amount of territory in both the Northern and Southern Hemispheres. High-altitude communications satellites can "see" a huge swath of Earth and so can send and receive signals to and from many Earth stations simultaneously. The portion of the Earth's surface that is illuminated by a satellite is called its "footprint." Except for the far north and the far south, one equatorial satellite can cover about one-third of the Earth's surface. Thus, theoretically, three geosynchronous satellites can cover most of the Earth's surface.

The cost of high-orbit satellites and transmission delays has caused some telecommunications providers to seek an alternative in smaller and cheaper satellites in low Earth and medium Earth orbit. Low-Earth-orbit (LEO) satellites constantly circle the Earth usually at an altitude of between four hundred and one thousand miles. Another orbit, called Medium Earth Orbit, or MEO, which is below geosynchronous orbit but above low Earth, can achieve global coverage with about 12 to 15 satellites. In addition to communications, MEO satellites are favored for global surveillance, environmental studies, and monitoring nuclear missile treaties. Commercial companies have now begun to launch constellations of LEO and MEO satellites, dozens or even hundreds of which will ultimately link together in a communications service web that surrounds the planet.

LEOs are deployed so that there is always at least one on the horizon to receive transmissions from an Earth station. A telephone call started with one overhead satellite is handed off to the next one that arrives overhead; conversations relayed by a LEO do not suffer from the delays experienced when using communications satellites in geosynchronous orbit. More of them are needed for full coverage of the Earth, but the failure of one satellite does not cause the failure of the entire communications system. The best-known LEO constellation is the sixty-six-satellite Iridium system. This $6 billion effort has begun to provide handheld satellite telephone services from any location on Earth to any other. Around the year 2003, Teledesic, an ambitious project to deploy 288 LEOs for high-speed video and data, is scheduled to start service. The project, in which Microsoft magnate Bill Gates is an investor, has sometimes been dubbed an "Internet in the Sky." The high cost of putting satellites into space has also spurred interest in unmanned High Altitude Long Endurance platforms that would hover above cities at about 65,000 to 100,000 feet, providing the capacity for two-way communications and distribution of video in a circle three hundred to five hundred miles wide. Power requirements are a possible limitation for this technology, though one design proposes the use of helium-filled dirigibles, while another would use very high efficiency jet engines.

## Wirelessing the World: The Cellular Telephone Boom

In 1895, twenty-year-old Guglielmo Marconi, heir of a well-to-do Italian family, built the first radio transmitter, sending a burst of radio waves to a simple receiver about one and a half miles distant. Thus was born a technology that allows astronomers to listen for signs of life in the universe, and permits sports fans to get the latest scores. It also underlies the cellular phones (also called mobile or radio phones) that are now a ubiquitous feature of technoculture in the United States and around the world. Indeed, wireless communications, spurred by several waves of innovation, are widely considered the fastest-growing segment of this global industry. In Finland, more people communicate by mobile phone than by fixed land line. Analysts project that by 2001, nearly half a billion people worldwide will subscribe to a wireless service of some kind.

With so many millions of people communicating on the move, one may wonder how cellular communications providers can give everyone a device to call any other phone, fixed or mobile, without ever having somebody intruding on their conversation. Indeed, until the 1980s, a mobile phone was an expensive and scarce commodity with long waiting lists because of frequency limitations; there was no way to share the limited range of frequencies available. Then came the concept of the cell, in which a geographic region is divided into sub-areas, each of which is equipped with a low-power transmitter. When making a call, a mobile phone first communicates with a base station in a cell that assigns it a given frequency and relays its call to the telephone network and then to the recipient. This method of apportioning the frequency spectrum among the users in the cell is called Frequency Division Multiple Access (or FDMA).

If somebody dials a mobile phone, the telephone network finds the nearest base station and transmits the signal. If a mobile-phone user starts to move out of the range of one base station, another base station picks up the signal so the callers can keep talking, a process known as a handoff. This system of dividing up service areas into cells, with each cell served by a base sta-

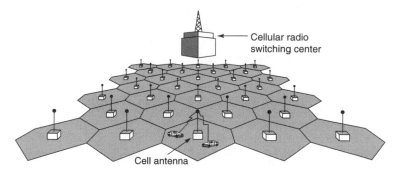

**Figure 3.5** Cellular telephone service providers divide a region into cells, each of which is equipped with its own antenna. Callers within a particular cell are assigned their own frequency so that there is no eavesdropping or interruption. As the user passes from one cell to another, the call is "handed off" to the cell and its frequency is changed.

tion, allows cellular service providers to reuse radio frequencies in the various cells of the network. The cell concept has been taken even further with PCS or Personal Communications Systems, a digital telephone network that relies on a multitude of low-powered antennas that are much cheaper than the large, expensive, high-powered antennas used in cellular service. (Some areas, of course, still lack any type of commercial cellular service.)

---

## SPECTRAL POLLUTION

The growing volume of mobile international communications traffic is exacting a toll on radio astronomers, who study the universe by collecting radio signals from the depths of space. Any object warmer than absolute zero (minus 459.67 degrees Fahrenheit) radiates electromagnetic energy. Because of their particular temperatures, many stars, galaxies, and interstellar dust and gas formations emit radio waves. This form of radiation is also the only kind that can penetrate the dust that pervades the universe. So scientists probing the universe gather radio waves for clues about its nature. Among the threats to radio astronomy are low-altitude satellites serving operators of paging devices, global cellular telephones, television broadcasts, and even the kind of radiation emitted by garage-door openers.

Radio astronomers fear that the Iridium network and other systems that use the same frequencies will seriously affect their work. The problem is that the sixty-six satellites relaying calls between distant points on Earth operate at microwave frequencies between 1610 and 1626.5 megahertz. These emissions will overlap a region of the radio spectrum between 1610.6 and 1613.8 megahertz, which radio astronomers regard as vital. Radio interference at these frequencies masks the very faint signals emitted by interstellar gas clouds and the extended envelopes of red giant stars. The inefficiencies of transmitters, which cause them to emit extraneous signals outside their designated frequency bands, can also interfere with radio telescopes. Iridium, in fact, has promised to turn the satellites off for selected periods as they pass over some radio telescopes. Another concern arises from a recommendation to the FCC by a group of paging-device businesses that asks that a portion of the radio spectrum between 1400 and 1427 megahertz be allocated to operators of low-altitude satellites serving pagers and other devices. This band overlaps the frequency of hydrogen, which makes up more than 90 percent of the matter in the universe. Some astronomers say that if commercial use of the hydrogen frequency is permitted, it will more or less put them out of business.

The first cellular networks were analog, transmitting voices by modulating radio frequencies. The developing standard in wireless communications is digital, and many of the original analog networks are now making the conversion. Compared to their analog counterparts, digital systems can both expand the capacity of the medium and compress the messages they carry. This packs more bits of conversations into a slice of spectrum than an analog system can. Digital signals are also harder to eavesdrop on. The digitization of cellular networks enables providers to deliver more services than simple voice transmission. Digital wireless networks harness "intelligent network" technology to locate and identify roaming subscribers outside their local calling area and to customize the services they receive.

One digital technique that permits multiple calls to take place within the same area is called Code Division Multiple Access, or CDMA. This system provides ten to twenty times the capacity of previous analog systems and perhaps two to three times that of the more conventional digital technique known as time-division multiple access. It had its genesis in military communications that had to be secure. CDMA assigns a unique code to every telephone call or data transmission. At the receiving end, this code distinguishes a given call from the multitude of calls transmitted simultaneously within the same band of frequencies. Only another CDMA phone that has been given the proper code can descramble an incoming call. CDMA is just one of several standards vying for customers as wireless networks go digital.

## WHERE IN THE WORLD . . .

Satellites have had a revolutionary impact beyond telecommunications. Finding one's exact location can be done with pinpoint accuracy through the NAVSTAR Global Positioning System and its network of nearly thirty satellites that occupy a medium Earth orbit about 12,500 miles above the Earth. These satellites, placed in orbit by the U.S. military, allow boats, airplanes, wilderness adventurers, and anyone else with the right equipment to check their location on Earth to within a few meters. In some luxury cars, a computer in the vehicle uses mapping software working in conjunction with a GPS unit to help find the best route to a designated location. A display shows an icon of one's car moving toward the destination address that has been entered. Wrong turns are instantly flagged.

The GPS system was constructed by the Department of Defense to allow military ships, aircraft, and ground vehicles to determine a location anywhere in the world. (The Russian Federation also has a satellite navigation system, which is called Glonass.) Although the designers of the GPS intended it for secret military use, they did make provisions for civilians to locate themselves, but not with total accuracy. Nominally, the diminished accuracy of the signal should provide readings that allow location of one's position to within 100 meters. Nevertheless, scientists and engineers working outside of the armed forces have devised ways to compensate for the purposeful degradation of the GPS signals, providing accuracy to within a few meters. Ordinary civilians can now achieve much better results than the U.S. military expected, and restricted access by the government may be phased out. Commercial applications now far outnumber the military uses of the system, and in fact, during the Gulf War in 1990, American soldiers finding their way through the featureless deserts of Kuwait actually used commercially available GPS systems to help them. This satellite positioning system has been placed on the top of Mount Everest to help geologists understand this earthquake-prone area. The GPS surveying equipment can measure shifts in the Earth's crust of just a few millimeters. Farmers utilize the equipment to survey the condition of small sections of their fields so that they can distribute fertilizer more effectively.

To determine a location, a receiver GPS measures distance from three or four satellites from the timing of pulses sent from these satellites. The navigation satellites transmit radio pulses at a specific, known time—measured by an atomic clock aboard the satellites. By measuring the exact instant when the pulses arrive, the receiving equipment determines the distance to the satellite. The system then uses these readings to pinpoint a location by using a principle called Spheres of Position. Each satellite orbits in a precisely known pattern. Thus the distance coincides with a set of possible locations that form an imaginary sphere, with the satellite emitting the signal at its center. The intersection of several spheres with the surface of the Earth marks the person's exact location.

## Shared and Separate:
## The Distributed World of Computer Networks

The technologies of telecommunications and computers are remarkably intertwined: modern telephony is saturated with microprocessors—from the computers that switch and route millions of telephone calls every day to the photoreceptors that receive and decode the on-off light signals of a digital data message that has traveled thousands of miles along a fiberoptic cable. Similarly, with the growth of distributed data processing—in which tasks for a given endeavor may be apportioned among different computers—the telecommunications network has become an extension of the internal memory, storage and communications that reside in each computer. Data that travel around computer networks can be routed along a company's private transmission lines or else sent on their way over the public communications lines.

The roots of distributed information processing go back thirty years, to the infancy of computer networking. To use the very expensive processing power and memory of mainframe computers, time-share systems were developed in which dumb terminals (which lacked processing power) were linked to mainframes that could be hundreds of miles away. The nature of networked computing changed, first with the ascent of the minicomputer, and the trend accentuated greatly with the arrival of the personal computer. Computing functions are no longer centralized in a mainframe; instead they are distributed among a number of personal computers or workstations in a network. These combine the processing power and memory of multiple computers, making it possible to achieve a combined performance that is comparable to that of a supercomputer. In the 1980s, local area networks (LANs) were developed to connect devices usually in a single building or group of buildings. The most common LAN is called Ethernet and was developed by Xerox, Digital Equipment, and Intel. In an Ethernet LAN, a computer connected to the network uses a protocol that "listens" for gaps in transmission before transmitting data to another computer over a variety of media, from optical fibers to telephone wiring.

Individual LANs operate in a limited area. But other networks may hook LANs together over greater distances using high-speed telephone lines and specialized hardware. A metropolitan area network (MAN), for instance, can tie together corporate campuses throughout a major urban area, while a wide-area network (WAN) might serve corporate offices to be connected throughout the country or worldwide.

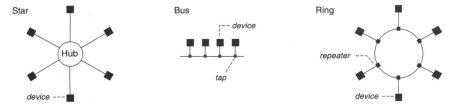

**Figure 3.6** Local area networks, or LANs, link together large numbers of computers and other peripheral devices that need to communicate with each other. The topology of a LAN describes how the devices are linked. Three common topologies are star, bus, and ring. In a star configuration, a device called a hub relays messages to other computers. A bus lets computers tap into a central connection. The ring circulates a message to a destination computer.

Computer networks are designed to allow data to travel among many users with little delay. The design of the network—the way one computer is connected to another—is called the *network topology*. One network topology might be described as a star in which a corporate mainframe, the centerpoint of the star, maintains links with, communicates with, and controls other devices on the network. Computers that store the information accessed over the network are called *servers*. And the user's computer and software that interact with a remote server are called a *client*. A Web browser, a type of client program, might request information about rheumatoid arthritis from a server at the National Institutes of Health, for instance.

## High-Speed Delivery: The Technology of Switching Calls

A physical transmission medium—whether cellular or fiberoptic—cannot simply transmit from one point to another. Switching equipment must route a phone call or data to its destination. Switching is analogous to routing trains along an intricate web of railroad tracks. With the number of telephones in the world exceeding 800 million, however, connecting a telephone (or modem) call with another anywhere in the world is an enormously complex undertaking.

In a telephone call, when the handset is lifted, an electric circuit is established, its power supplied by the nearest phone office or switching station. This circuit signals a station to send a dial tone, indicating that the telephone or modem is now wired into the local, regional, national, and global network. The local station relays the connection to another telephone if the call is local, or to an outside trunk line that will send the call to the next station for a long-distance connection.

From the advent of telephony in the late 1800s (when switchboard operators manually connected calls by literally plugging the two callers into a circuit) until the 1970s, the automatic exchange switching systems of the world were electromechanical. Those exchanges used motors and electrically powered relays to establish a connection between an incoming voice channel and a desired outgoing circuit. The connections were slow-moving and the switches required considerable maintenance.

With the invention of the transistor, fast and reliable digital switches usurped electromechanical switching. State-of-the-art switching is done by powerful computers and mammoth software programs. For example, one of the 5ESS switches, developed by AT&T ( now Lucent Technologies), handles more than 190 million calls on a normal weekday and is controlled by a program comprising about 2 million lines of code and costing about a billion dollars to develop. Like other digital technologies, the increasing speed and miniaturization of microelectronic circuits have led to the development of faster switches with enormous capacity. A switch recently developed by the giant Siemens corporation can handle more than a trillion bits per second of digital information, which the company claims is the equivalent of all the telephone calls of the world, in one control board.

Certain types of switches can also help transmit computer data efficiently, avoiding the need to hold a line or signal path open during a phone call or transmission, thus tying up network capacity whether or not data is being transmitted. A technique called *packet switching* splits up a stream of data into short segments or packets, which might be thought of as tiny data-carrying boxcars. Each packet has a header containing information about the identity of the message, the sender and receiver, the sequence of packets in a given message, and the message priority. Packets are then split up and sent over whichever lines are most available; a message is then reassembled at the receiving computer.

Though switches today already represent enormous advances over their electromechanical forebears, their limitations constitute the main obstacle to multimedia networks that integrate all kinds of information, including voice and video. Because older switching technologies are not capable of handling different types of traffic, especially the torrent of bits needed to transmit high-resolution video, networks have been unable to transmit the kind of live video that would make teleconferencing an everyday way of doing business. To increase the capacity of switches, networks have adopted what is called "fast packet switching," which allows the flood of packets from high-bandwidth networks to be switched rapidly. The most notable of the fast packet switching concepts is called Asynchronous Transfer Mode, or ATM. It is a protocol for transferring multimedia—voice, video, audio, or data—by placing digital information in short packets that are reassembled at their destination into a cohesive message. Switches that use the ATM protocol have begun to be used in recent years by telecommunications providers and by some businesses on their internal private networks.

Other devices that help direct the flow of information on large networks are called *routers*. Routers use complicated programming instructions and a considerable amount of processing power to connect different networks and to determine the most economical and fastest path from a sending to a receiving computer along the web of connections that make up the Internet. A message will traverse a series of routers—as well as some switches—before arriving at its destination. The job of the routers is to ensure that the packets always arrive at the proper destination. Besides routers, another hardware (and software) device that processes packets and routes them to their destination, mostly on private networks, is a *hub*. Hubs link groups of computers to one another and often let computers communicate as if they were a telephone party line, with all data flowing by every computer connected to the device.

## From Porn to Global Conversations: The Impact of the Internet

Perhaps the most prominent symbol of the interconnection of global computing and communications systems is the Internet—the collection of networks that links millions of computers around the world. Though it was decades in development, the "Net" has experienced its most explosive growth in the last few years: in the fall of 1990 there were just a few hundred thousand computers on the Internet; as of 1998, the number was in the tens of millions and growing.

The Internet has become such a presence that it seems that virtually every major organization and home has a Web address and a "home page" to visit. This impression is somewhat misleading. In fact, as of 1998, less than one-third of U.S. households were connected online, having grown from just 7 percent just a few years ago. But these figures must still be compared with a medium like television, with its virtually total penetration of American homes. And the numbers drop substantially for the rest of the world. But the Internet's impact on society is so great that it is sometimes difficult to understand the network's physical structure—or to conceive of how it came about.

The Internet consists of tens of thousands of networks of computers interconnected with each other over the international telecommunications system. A computer based in Johannesburg, South Africa, for example, is connected by modem to a local *node* or connection point that is itself connected to networks around the country and overseas. An E-mail message can be sent for only the charge of a local phone call. Internet providers whose nodes provide links to the worldwide network do charge a fee, which is relatively low in the United States but can still be a hefty amount elsewhere in the world.

No single person, group, or organization runs the Internet. This seems wholly appropriate for a multitude of networks that interact with other networks to pass messages. The Internet Society and the Internet Engineering Task Force (IETF) oversee the daunting task of maintaining the well-being and growth of the network. The IETF is made up of volunteer technical professionals who establish standards and protocols, monitor the running of the network, and plan for the future. In mid-1998, moreover, an effort was under way to establish a new nonprofit corporation to coordinate how Internet addresses are assigned and to handle other network-related administration functions. Yet another organization, the World Wide Web Consortium (W3C), develops standards for the evolution of the fastest-growing part of the Internet, the World Wide Web. The Web consists of servers connected over the Internet that store documents that may contain text, graphics, audio, and video that are accessed with software called a browser. Run by the Laboratory for Computer Science at the Massachusetts Institute of Technology, W3C serves as a storehouse of information about the Web for developers and users.

Considering that the Internet has become a repository of information that ranges from DNA sequences of the human genome to advertising for catalog shoppers, one might be surprised to learn that this global network was fashioned by the United States military. In fact, the Advanced Research Projects Agency, a research arm of the Department of Defense, coordinated the building of the early Internet. ARPA's network engineers faced the task of designing a robust network that could share the resources of scarce, expensive, and incompatible mainframe and minicomputers among groups of different scientists.

Thus, in 1969, ARPANET, the first ARPA-sponsored computer network, went into operation. The developers of the Internet soon created protocols that allowed different types of computers or even distinct networks to "talk" with one another. They devised what became known as the Transmission Control Protocol/Internet Protocol or TCP/IP, which remains the de-facto set of protocols used by computer networks around the world. The creation of ARPANET required network engineers to solve a daunting series of technical problems: connecting computers without requiring every computer to be linked directly to the other and hooking together computer systems from different manufacturers that had many varieties of operating software.

The solution was what is known as *store and forward packet switching* (see page 71): Instead of having every computer linked to every other, store and forward technology routes messages through the network. Backbone communications lines, high-speed links that carry most of the traffic, connect networks or computers together in the same way that the spinal column serves as the conduit for the myriad nerve impulses that travel through the human body. These backbones used to be run by the National Science Foundation, but are now paid for by government agencies and by private corporations, which provide Internet access to customers. The computers, or nodes, that act as the switching centers receive packets and pass them along to the next node. A computer at the message's destination reassembles the original message from the myriad packets.

Information sent across the Internet is first broken up into packets by the Transmission Control Protocol, which runs in software on a local computer. The packets move from a computer to a local network, Internet service provider, or online service. From there they are sent through many levels of networks, computers, and communications lines before they reach their final destination. On their journey they are processed and relayed on their way by a variety of hardware, such as devices called routers (see page 71). If the destination of the packets lies outside a set of intermediate networks, they are sent to a NAP (network access point) to be dispatched across the country or the world on a backbone.

## The Journey of an E-Mail Message

One of the most helpful ways to understand the structure of the Net is to analyze the syntax of an E-mail address. Like an ordinary letter, electronic mail requires an address. Every address on the Internet is actually a series of numbers separated by periods, which are called dots. Because these numbers can change and are difficult to remember, a so-called Domain Name System (DNS) is responsible for tracking these addresses using letters and words that are easier to remember. Domains are groups of computers that are identified by names such as ".com" (pronounced "dot-com").

Take the hypothetical journey of an electronic mail message sent to John Smith at Acme Computing Company. The sender deploys a client, or software program, to compose a message, and affixes the electronic mail address, *john.smith@acme.com*. Operating system software, called Transmission Control Protocol (TCP), divides a message into packets and attaches information about how each packet should be handled by the network. The message is usually passed from the sender's computer to a server, a computer connected to the Internet. After consulting computers on the Internet that supply information about network addresses, the server converts the domain names in the recipient's electronic mail address into a numeric Internet Protocol (IP) address. It reads the domain name from right to left: ".com" for "commercial site" and "acme" for the destination computer server at the Acme Company. The IP address enables the packets to move through ten or so "router" computers dispersed throughout the network before arriving at Acme's mail server computer. The mail server stores the message in the "john.smith" mailbox. Then Smith's computer can retrieve the message.

Today the Internet is widely used for electronic mail. But it also includes many other services. Usenet is the equivalent of a global soap box. It consists of tens of thousands of special-interest areas called *newsgroups*.

A Usenet newsgroup allows messages to be posted where they can be seen and commented on by others. Usenet groups range in topic from art history to religion, politics, sex, cooking, sports, and warfare. The system emerged in 1979, the brainchild of students at Duke University and the University of North Carolina who developed a system known as Unix-to-Unix CoPy (UUCP), which allowed users to automatically download text or software from Usenet groups. New versions are now widely used for managing files by other Internet services.

## Exploring the Net: The World Wide Web

Many people's notion of the Internet is synonymous with the World Wide Web. The Web, a hypertext system, was developed by the CERN High Energy Physics Laboratory in Geneva in 1989. It has been noted that Switzerland, where four languages are spoken, including German, French, and Italian, was an appropriate place to develop what has become, in essence, a common language for a worldwide communications system. Tim Berners-Lee, the inventor of the Web, reportedly envisioned it as a growing superhuman "brain" formed by linking together people's knowledge around the world.

The Web became possible because of a series of software innovations that made feasible the concept of networked hypertext, allowing users to jump from one site to another. Berners-Lee came up with an addressing scheme called the Universal Resource Locator, or URL (pronounced "earl"), for locating files, pictures, audio, and video anywhere on the Internet. An example of an URL on the World Wide Web is *http://www.wiley.com*, which is the URL for John Wiley & Sons, Inc., the publisher of this book.

Berners-Lee also devised a simple programming language called hypertext markup language (HTML) for creating "pages" of information that might contain, say, a video animation clip with accompanying text displayed beside the image, similar to a newspaper layout. Browser software that runs on a user's computer accesses the remote computers where text, objects, and pictures are stored using the hypertext transport protocol (http). The protocol defines how Web browsers request and obtain Web pages. To refer readers of a Web page to text, images, or other media on the Web, a reference is underlined and programmed with the URL (or hyperlink) of the site to be accessed. Likewise, other sites can include hyperlinks to the same page. This virtually infinite capacity for cross-linking using simple rules of association is a key reason the World Wide Web is such a powerful, innovative publishing medium.

Because there is no way of stopping anyone from publishing anything and placing it on the Net, the Internet has started to represent a vast, disorganized library that contains millions of poorly cataloged documents. No one ensures that information is accurate or up to date. Also, despite vast stores of medical, financial, and other useful technical information, the Internet is also a repository for schlock, including pornography and thousands of personal Web pages that shower the visitor with the site owner's pet obsessions.

Bringing order and stability to this universe of information has now become a major research issue. Software that helps track down information on the Web is called a search engine. With names like HotBot and Alta Vista, search engines crawl day and night on the Web. They are programmed to store every significant word they find in a gigantic index. Once the index is compiled, a computer user need only type in a key word or phrase, and the software will rapidly list pages that have been indexed with matching words. This kind of service can often present researchers with the daunting task of sorting their way through thousands of "hits" that list any document with even a passing reference to a topic.

Moreover, despite their assiduous journey throughout the Web, most of the Web still remains uncataloged. Besides searching the Web,

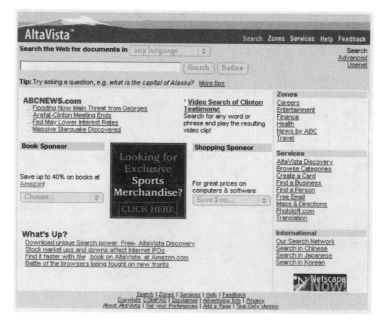

**Figure 3.7**  A search engine like Alta Vista continuously trawls the World Wide Web to compile indexes that help befuddled Web travelers find their way.

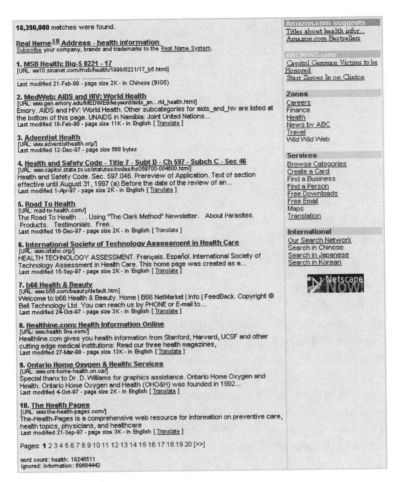

**Figure 3.8** A query on health information using Alta Vista rapidly produces a list of Web sites.

browsers sometimes contain filtering software that can be used to cull the good from the bad. It can prevent minors from gaining access to pornography or other offensive sites. But it also may flag desirable sites—those that provide quality information on health, for instance. Standards groups have labored to create a rating system that can be used in conjunction with software filters, allowing a browser to pull up sites that have attained a certain ranking.

But domesticating cyberspace remains problematic. Many other issues that are normally handled through government regulation—e.g., whether Web sites have the right to collect information about their visitors and whether Internet usage can be taxed—still remain hotly contested.

## The Ultimate Long-Distance Bargain: Making an Internet Call

Because of the structure of the Internet, corresponding with an overseas friend by E-mail is considerably cheaper—and faster—than mailing a letter. The efficiencies of the Internet's packet communications have not gone unnoticed by entrepreneurial companies who have started to specialize in voice calls over the Internet. Internet telephony is an evolving technology. In the mid-1990s it required a fast personal computer equipped with a speaker, a sound card, and a headset as well as a high-speed modem with a direct connection to the Internet. The user would speak into the computer microphone, and special software would turn the voice conversation into digital "packets." The message could then be relayed to someone else who had purchased similar equipment. But the data packets that make up a phone call could get backed up behind those from someone downloading a pornographic picture. So the conversations were often fuzzy-sounding.

Internet telephony has begun to target a wider audience. Improvements in communications software have resulted in protocols that let voice packets be delivered before data packets, which has meant better-quality phone conversations. And some Internet phone companies compete with the likes of AT&T and MCI by offering service using regular phone sets. Many of the largest carriers have themselves begun to make plans to deliver phone service using packet networks. Internet calls are cheaper, in part, because they use bandwidth more efficiently. The calls are diced up into packets, compressed, and then made to fill a communications pipe from end to end. In ordinary telephone connections, bandwidth is wasted by the pauses and silences during a conversation.

## The Perils of Distributed Computing: Encryption and Data Security

Distributed computing—in which various processing tasks are performed by different computers—coupled with the growth of the Internet, has taken enormous amounts of information out of computer centers and made it readily accessible to anyone with a computer and a modem. This development has raised serious issues of data security. Such networks are vulnerable to intruders, or hackers, who take pride in breaking into computer networks sometimes just for fun, and on other occasions stealing proprietary information, credit card numbers, or sensitive military information. The Internet is particularly vulnerable: sending a message over the Net has been compared to mailing a postcard that can be read by anyone. In the spring of 1998 a group of hackers who identified themselves to a Senate committee as "Mudge," "Space Rogue," and "Brian Oblivion" claimed that they could cripple the Internet in half an hour. Whether such boasts are overstated or not, resolving security and privacy issues are critical to changing perceptions of cyberspace as an electronic version of the Old West.

Corporations protect their internal data networks from outside predations by setting up *firewalls*—combinations of software and hardware that prevent outside access by examining the contents of Internet data packets addressed to computers on a corporate network. But firewalls are not totally impenetrable, so some organizations that transmit sensitive information have chosen to forgo using the Internet entirely.

Security and privacy concerns remain a critical barrier to making the Internet a medium for round-the-clock shopping. Both vendors and buyers are concerned about the security of sensitive information like credit card numbers, and many online retailers now use systems that routinely encrypt any information a customer enters to order a product. Encryption involves encoding the data so that no eavesdropper or thief can read the contents in transit. One encryption program that is being readied is the Secure Electronic Transaction Protocol (SET), a group of procedures and protocols designed to make financial transactions on the Internet as confidential as possible.

To protect sensitive information—whether a credit card number, a password, or a sensitive military message—software engineers design complicated codes. Breaking those codes often involves solving an extraordinarily difficult mathematical problem. For example, the solution to a code might be to find two numbers that can be multiplied together to generate a number that is several hundred digits long. This process, which is called factoring, is currently considered beyond the capability of even the most skilled hacker.

But if you are sending encrypted data, the recipient needs to know how to decode the information. Networks that send and receive encrypted information usually use *keys*: a string of numbers that can be used in a series of mathematical operations to scramble or unscramble the digits representing a message. A method called public-key cryptography avoids the problem of requiring both the sender and the receiver of a message to possess the same secret key, which risks being stolen. Public key cryptography allows a message to be encrypted with a publicly available key that can only be decoded with a private key held by the recipient. The best known public-key system is RSA. Public-key cryptography will serve as a foundation for electronic cash, digital personal signatures (which verify identity), and copyright permission schemes.

The issue of computer network security will continue to preoccupy not only technical mavens but policymakers as well. It is considered so sensitive that the United States government has long restricted American software companies from exporting certain encryption software on national security grounds. Though breaking these codes is well beyond the capabilities of the average hacker, computer scientists understand the vulnerability of algorithms previously thought to be too difficult to decrypt without enormous expenditures of time and money. In 1998, for instance, a group of researchers used a custom-built computer, which cost less than $250,000, to break a message encrypted with the U.S. government's Data Encryption Standard in less than three days.

Work also continues on hypothetically unbreakable schemes. One intensively researched method builds on principles of quantum mechanics, which stipulate that the mere act of observation can change the state of an electron or other basic constituents of matter at the atomic scale. In the case of quantum cryptography, digital information can be transmitted as particles of light, or photons, that are oriented (polarized) in, say, a horizontal or vertical direction. The polarization of each photon represents the "zero" or "one" state of a digital bit. If a hacker tries to intercept the transmission, this attempt to ascertain the numbers that make up the key will cause some of this information to be lost, an indication of tampering. Quantum cryptography might be used to ensure that a secret key sent to someone for decoding an encrypted message has not been hijacked.

Another area related to computer security that has captured the public's attention is the transmission of computer viruses. Often lurking in files accessible on the Internet—as well as on floppy disks and CD-ROMs—a virus can "infect" other programs or files surreptitiously and begin to multiply until it destroys data or disables a program. Thousands of viruses have appeared in recent years. And an industry has developed to "innoculate" computers against viruses. Some of these anti-virus programs are even modeled on the human immune system. However, the threat of viruses is sometimes exaggerated, and many virus hoaxes have been promulgated.

## Nets of the Future: The Coming Information Avalanche

Within the past decade, the Internet has developed into a stunning cornucopia of international communications and data resources. But in its present state, the ease of use of the network of networks is limited in part by the capacities and speed of its links and switches. This is clear to anyone who explores the Internet using a modem that sends and downloads information at speeds of 28,000 bits per second. It can take twenty minutes for the transfer of a short video clip or a sample of music. Additionally, using hypertext links to explore the World Wide Web is often an exercise in tedium as the user is forced to call up one short link after the other. Nevertheless, given the growth of the Internet, and its penetration of everyday commerce and culture, there is no doubt that its present incarnation is but a harbinger of its future potential.

Even more advanced communications are on their way as broadband networks and switches are implemented and manufacturers and governments worldwide agree to standards for linking the diverse components of global networks into a seamless web. The U.S. government has sponsored a project, dubbed Internet 2, that will link one hundred universities, allowing transfer of data at speeds of more than a billion bits per second.

Within the next decade, techno-pundits predict a halcyon future in which the World Wide Web will reach into virtually every home, school, and business with high-speed links that feature high-resolution sound and video. Schools of the future will have ready access online to the same resources as scholars and scientists, and the whole record of human history will be deposited and accessed from virtual World Wide libraries. Increasingly sophisticated search engines will make it easy to find relevant information. And personal Internet secretaries (known as intelligent agents) will observe someone's interests and then seek out and deliver this information. Virtual corporations will consist of widely dispersed individuals and groups who work together for a time before moving onto the next project with a wholly new set of collaborators. Their only common bond: the ability to communicate over electronic networks.

While much Internet information is provided for free—or is sponsored through advertising—you pay for what you get: those who provide quality information and entertainment will need some form of remuneration. Encrypted cash and identification will provide the technological means to carry out such transactions. Bank notes may gradually be replaced by digital cash in systems that maintain account balances and credit information online. Digital cash will be transferred from an online account to pay for a garment ordered from an electronic retailer. Digital certificates will verify the identity of a consumer to an online merchant and also confirm who the merchant is to the consumer.

Like a wallet or a keychain, networked computing may become coupled to one's personal possessions. Digital personal communications systems the size of a wristwatch are already available for making phone calls. A plethora of other such systems, including wearable computers, active badges, and personal navigation devices, will become increasingly prevalent during the coming decade. These devices consist of small computing units and transmitter-receivers that process a signal to allow the user to interact with his or her surroundings. An active badge, for instance, may transmit a communication signal to indicate the whereabouts of an employee. In fact, prototypes of badges let people exchange information about each other—name, title, telephone, and electronic mail—by an infrared link across a room.

The personalization of the computer may be taken to its extreme with the BodyNet, also called the "personal area network," which is intended to function as an extension of the person's body. Such computing devices may be strapped to oneself to exchange information or to monitor changing bodily conditions. They may prove especially useful on the factory floor or in other settings where the user needs both hands free.

"Smart" eyeglasses consist of a wearable computer with a head-mounted display, cameras, and wireless communications. One MIT researcher used a BodyNet to relay to his wife at home an image of supermarket produce. She then sent back an electronic mail message about her choices, which was displayed on the eyepiece of her husband's BodyNet. The ultimate in computer-corporeal interaction may be smart underwear.

**Figure 3.9**  In the future, wearable computers may be embedded in your underwear, but for now, they do not make a fashion statement. In this picture, researchers at the MIT Media Lab display their wearable gear.

A sensor in an undergarment can relay the level of perspiration and then adjust a heating unit accordingly. But any benefits from adjusting a thermostat by a touch of perspiration may be overcome by the discomfort of loading underwear with electronic sensors. Even the inventors of these devices recognize some of the drawbacks. One researcher raised the issue of what would happen if one person begins to sweat before another. "Multiple users create some interesting problems. But a discussion of how, for example, husband and wife mutually control the temperature of a heater is a subject for a marriage counselor," he said.

Smart underwear is supplemented by visions of a digital future in which people "meet" and interact in real time with a friend on another continent or live in a house that instantaneously adjusts temperature and humidity as weather conditions change. The interaction may occur, however, between two virtual entities. A person may be represented in cyberspace by an *avatar*—a software alterego that may or may not bear a resemblance to someone's actual physical presence. The smart home also looms large on the digital horizon. Computer and electrical products companies have tried to hammer out standards that would enable a home's electrical power lines to transmit data that could control home appliances, lighting, temperature, alarm, and sprinkler systems. Or these signals might otherwise pass on wireless links.

This Jetsons-like future may be farther away than some visionaries predict because of limitations in the key technologies. Few purchasers may step forward if the control system for home heating and air-conditioning system prove as troublesome as a personal computer operating system. The wired home and BodyNets may be cumbersome or costly and, for the time being, may be embraced only by the very rich or by those enamored of technology for its own sake. Microsoft founder Bill Gates's $30-million Seattle home can display pictures and music to suit the tastes of the occupant of a room after being cued by an active badge. But after the novelty of such a device wears off, few homeowners are likely to pull themselves away from meal preparation to program badges for their dinner guests.

Given the enormous infrastructure investments required by communications technologies, there are concerns, as with other technologies such as genetic engineering, that wiring the world will benefit rich nations more than poor ones, and widen the gap between the haves and the have-nots. Social critics worry about threats to personal privacy—as a networked world allows the government more easily to compile information on its citizens. Indeed, users of a recently implemented system to automate road-use tolls are concerned that information generated by payments at varying locations may be used by law-enforcement agencies to track a person's comings and goings.

The costs must be weighed against the knowledge that networks can potentially deliver enormous advantages for social development. A phone line is needed for any Internet connection. And in the poorest countries, the average density of the phone lines is about five per one hundred people. In some cases, however, mobile telephone technology let some of these countries "leapfrog" the old copper network by delivering telephone service to remote areas without installing costly and time-consuming wiring infrastructure. The growth of the Internet, moreover, has begun to link remote corners of the world. When President Clinton visited Africa in April 1998, readers in the United States could follow local perspectives by accessing newspapers in Ghana and South Africa over the Internet. With higher-speed data links, the emerging field of telemedicine (see chapter 7) would make it possible for doctors in sophisticated Western hospitals to make diagnoses on patients in less-developed countries oceans away. For example, a Boston-based physician might might view a satellite-relayed video image of a heart patient in Thailand while manipulating a surgical instrument to cut or seal blood vessels from his or her office. The challenge, as in other rapidly developing technologies, is to make wise use of these capabilities.

## TIMELINE

The construction of what is rapidly becoming a worldwide network of communications services is frequently described as being the world's most costly and important technical achievement. This short chronology identifies some key developments that were the intellectual and technological precursors of modern communications.

**1770s** The semaphore, the first practical system of long-distance telegraphic communication, uses flags held in different positions, or flashing lights mounted in towers, to convey messages letter by letter. By 1844 the French network connects twenty-nine cities through about five hundred stations over 4,800 kilometers and employs 3,000 operators and workers. The semaphore may be regarded as a precursor to present-day microwave radio-relay systems, which often use hilltop sites.

**1837** The electric telegraph is patented in the United Kingdom. Inspired mainly by the need of the expanding railway systems for a fast and reliable means of communication, the first system uses five magnetic needles, each of which can be detected by an electric current. By deflecting the needles successively in pairs, any one of twenty letters in the alphabet can be selected in turn.

1835    Samuel F. B. Morse invents a "printing telegraph" in which a key can switch on the current of an electromagnet as long as the key is held down, thereby causing a pencil to make long or short marks on a moving strip of paper. Morse's key invention was the code that bears his name. Morse code assigns different combinations of dots and dashes to the letters of the alphabet and numerals. Inherent in Morse code is the concept of transmitting information by coded groups of on/off signals.

1843    The electromechanical fax machine is patented by the Scottish inventor Alexander Bain.

1853    A telegraph wire sends two messages simultaneously, one in each direction.

1874    Thomas Alva Edison invents a quadruplex circuit, enabling the simultaneous transmission of four messages on a single wire.

1866    The transatlantic telegraph cable goes into service. By 1875 a worldwide network of telegraphic cables is laid.

1876    Alexander Graham Bell patents the first telephone. The head of Western Union, presiding at the time over a telegraph system that is the largest company on the planet, calls Bell's invention "an electric toy."

1889    An automatic telephone switching system debuts.

1897    Guglielmo Marconi applies for a patent for wireless telegraphy.

1907    The thermionic—or vacuum—valve is invented. This valve can amplify, generate, and modulate carrier waves. Such valves, or tubes, will be the standard until the invention of the transistor.

1913    AT&T, under Theodore Vail, initiates the goal of supplying service to the public at large—a policy called universal service.

1920s   The era of long-distance short-wave communications begins.

1937    Pulse code modulation is invented. This technique enables the transmission of digitally encoded signals over long distances without introducing noise, distortion, or loss of signal strength.

1945    Frequency modulation, a method of transmitting radio signals that avoids static and interference, becomes widespread in the United States.

1948    The transistor is invented.

1950    The junction transistor makes its first appearance in a telecommunications transmission system.

1956    The first transatlantic multichannel submarine cable system is laid, providing thirty-five circuits for commercial use.

1957    The Soviet Union launches Sputnik 1; signals from its VH1 transmitter were received by radio amateurs and others throughout the world.

1960    The Echo 1 communications satellite is successfully launched, and is used to test the efficacy of microwaves for communications purposes.

1961    The Bell System installs the first digital carrier T1 system—allowing transmission of 1.54 megabits per second.

1962    Telstar is launched—the first commercial satellite to provide active relay of live television and voice communications between the United States and Europe.

1967    Theodore Nelson describes hypertext—notes, documents, and other media digitally linked to one another. A quarter-century later, hypertext becomes the primary user interface of the World Wide Web.

1968    The U.S. Department of Defense's Advanced Research Projects Agency funds the ARPANET project to network four campus computer centers together.

1970    AT&T lays an eight-hundred-circuit cable between the United States and Spain.

1975    Field trials begin of an optical fiber telecommunications system; by 1983 an intercity trunk system is installed between Washington, D.C., and New York City.

1988    A transatlantic optical fiber cable system is laid between Europe and North America that has a capacity of 40,000 simultaneous telephone conversations and is capable of transmitting data and video signals.

1990–91  World Wide Web software and protocols are developed by the CERN high-energy physics laboratory in Geneva.

1996    The Telecommunications Reform Act of 1996 opens the new media floodgates in the United States by laying the groundwork for companies to compete in formerly regulated local markets.

1998    The Iridium network of sixty-six low-Earth-orbit satellites, which undertakes to provide worldwide telecommunications coverage, goes into operation.

# TALKING TELCOMM

**analog** Information, like human speech or music, that consists of continuous and smoothly varying waves over a certain range.

**bandwidth** The range of electrical frequencies or the amount of digital information a device or line can transmit in a given amount of time. Sometimes bandwidth is expressed as bits per second that a circuit or network can transmit.

**baud** A unit of signaling speed equal to the number of signal events per second.

**bit** A contraction of "binary digit." The smallest unit of information a computer processes. It can represent on or off, high or low, yes or no, one or zero.

**carrier signal** A single frequency that is modulated by a modem or other device to carry information.

**central office** A local telephone switching center that provides connections and services, such as voice mail or "call waiting," to subscribers or telephones within a particular area.

**data compression** A process that eliminates or reduces empty spaces or redundant information to pack more bits in a communications line.

**duplexing** Two messages transmitted simultaneously from opposite ends of the same circuit.

**electromagnetic spectrum** The range of frequencies, extending from very-low-frequency radio waves to light waves and ultimately to the highest-energy cosmic waves.

**ESS (Electronic Switching System)** A computer-controlled information connection or distribution system.

**fiber optics** Transmission of infrared light frequencies through a low-loss glass fiber with a transmitting laser or light-emitting diode (LED).

**frequency modulation (FM)** A method of sending information in patterns of frequency variations over a carrier signal.

**frequency division multiplexing** Communication system serving several users, in which each is assigned a different frequency to transmit data.

**ISDN (Integrated Services Digital Network)** Designed to be the digital replacement for conventional analog telephone service, ISDN uses the existing infrastructure of standard copper phone lines, but offers new capabilities such as simultaneous transmission of voice and data.

**microwaves** Frequencies in the electromagnetic spectrum between infrared and short-wave wavelengths.

**modulation**   Control of the properties of a carrier signal so that it contains the information patterns to be transmitted.

**multimedia**   Software and hardware that accommodates text pictures, sound, and video in the same application.

**packet switching**   A technique that allows many users to share a communications line simultaneously, in which data is shuttled around a network in packets, so that a single user does not monopolize a line. A packet is a fragment of a message that also contains information about the sender and the destination. A long message is broken up into a stream of packets, which are sent individually over different pathways in the network. The computer at the destination is responsible for reconstituting the original message from the received packets.

**protocols**   Software or hardware that manages the transfer of information.

**routers**   Switches that direct packets of information to the next link on the network—another router or the computer that is the packets' ultimate destination.

**time division multiplexing**   Sending information during an allotted time segment. A user transmits at assigned intervals; information is combined with that of users transmitting in other time slots.

**wavelength**   The distance between successive "peaks" or "troughs" of a wave.

# 4

# *Lasers:*
# *The Light Fantastic*

One does not usually associate Albert Einstein with compact disks and supermarket scanners. But in fact we owe the emergence of the laser to Einstein's inquiries into quantum physics and the nature of matter and energy. The laser, whose name is actually an acronym for "light amplification by stimulated emission of radiation," can be found in an array of devices whose common feature is a concentrated beam of pure, monochromatic light. In developing lasers, scientists have harnessed an exotic property of light straight out of quantum physics. And, as the centerpiece of science-fiction warfare, optical communications, and the mundane bar-code scanner, lasers epitomize the union between sophisticated scientific knowledge and human ingenuity that characterizes high technology.

Developed around the time of the first manned space shots of the early 1960s, lasers entered the popular imagination via the James Bond movie *Goldfinger*, which features a huge, hypodermic-shaped machine emitting a hot blue beam that nearly slices the hero in two. This appeared to be a confirmation of earlier predictions of death beams and ray guns, which dated back to the late-nineteenth-century H. G. Wells classic *The War of the Worlds*. Models in the General Motors "Futurama" exhibit at the 1964 New York World's Fair showed truck-sized lasers leveling trees in lumbering operations. Early laser scientists did little to discourage this macho image. They measured the power of their devices in "Gillettes"— the number of razor blades a beam could pierce. They had tapped a force that seemed as potent as nuclear fission.

Since then, the history of the laser has been one of practicality winning out over machismo, with surprising twists and turns. Lasers, as the means to destroy incoming enemy warheads, were the linchpin of the Strategic Defense Initiative, called "Star Wars," championed by President Ronald Reagan in the 1980s. More generally, however, lasers have become a benign household workhorse, laboring innocuously in devices as diverse as CD players, computer printers, supermarket price scanners, and fiberoptic data communications equipment (see chapter 3). With dreams of Star Wars on the wane, the laser's most profound applications might emerge in the realm of the microscopic, where beams of laser light are already being used as "scalpels" and "tweezers" to grasp individual molecules and perform surgery one cell at a time.

## The Principle Behind the Laser: Technology Plays Catch-up

The origin of the laser dates back to conjectures Einstein made in 1916 about the nature of matter and light and the possibility of prodding an electron to release a new photon of light if its energy levels are appropriately increased. Einstein's predicted phenomenon, called *stimulated emission,* was confirmed in 1928. However, it would take scientists another thirty years to figure out how to harness quantum theory to produce pure beams of light, and a working laser would not be built until 1961.

Einstein, elaborating on work by the physicist Max Planck, was analyzing the way electrons interact with photons. Quantum theory describes energy as traveling in small packets, or *quanta,* which have the properties of both particles and waves. The theory dictates that electrons, orbiting the nucleus of an atom, have certain specific energy levels, depending upon the orbit they are in. Further, the electrons can only be moved from one level of orbit to the next by either absorbing or emitting exact amounts of energy. This energy would be a photon (a quantum of light) of a specific wavelength.

An electron can be bumped into a higher energy state if the correctly powered photon comes along. After a few billionths of a second, this electron will drop back to its original energy level and in the process emit another photon identical in wavelength to the one it absorbed. The phenomenon of an electron dropping down to a lower energy state and in the process giving off a photon is called *spontaneous emission.* Einstein's conjecture was that if another photon, with the same wavelength as the first, struck an atom while its electron was already in the higher, excited state, the electron would emit a new photon. Normally, any photons caused by stimulated emission are quickly sucked up by atoms with electrons in a lower energy state. If the surrounding atoms have also been stimulated into higher energy states, however, these, too, will produce photons through stimulated emission, and more and more light will be produced. The effect of photons making new photons will cascade and a fountain of light will burst forth.

Not just any light. Photons emitted through stimulated emission emerge identical in all characteristics. This gives the light some very useful—and unusual—properties. It will not only be monochromatic (all the same wavelength, and thus the same color), but also coherent (i.e., all the crests and valleys of the light waves will be aligned). By contrast, a conventional light source produces a jumble of light waves from its numerous individual atoms. The phase of the wave emitted by one atom has no relation to the phase emitted by any other atom, so that the overall phase of the light fluctuates randomly from moment to moment and place to place. This lack of correlation is called incoherence. (For a discussion of light waves and frequencies, see page 57.)

## Pumping Up the Laser

The key to getting the laser process going is raising the total energy level of the light-emitting substance. To produce that cascade of stimulated light, scientists need to boost the energy level of most of the atoms in a sample. The goal is to create a "population inversion" of energy levels. Adding the energy to do this is called "pumping."

The first device that utilized pumping to produce pure electromagnetic radiation was invented in the 1950s by a physicist, Charles H. Townes, who used electricity to pump up the energy level of electrons in the nitrogen and hydrogen atoms in a tube of ammonia, and produced a pure beam of microwaves. It took him three years to get this device—which he called a *maser*—an acronym for "microwave amplification by stimulated emission of radiation"—to work.

In 1957, Townes began designing a similar device that would emit visible light, an "optical maser." Because visible light is shorter in wavelength

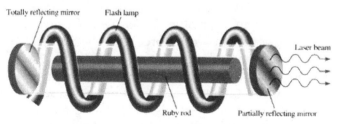

**Figure 4.1** A diagrammatic representation of the first ruby laser. The inventor, Theodore Maiman, used a strobe flash lamp to excite the ruby core. A totally reflective silver mirror amplified the stimulated light at one end, while the partially reflective mirror permitted the laser beam to escape.

than microwaves, more energy would have to be pumped into a medium to create the necessary "inverted population" of energy levels. Scientists everywhere began working on the problem. In 1960, Theodore Maiman got a cylinder of fabricated ruby crystal one and a half inches long and one-half inch in diameter to emit a 10,000-watt flash of brilliant red light. Lasting 300 millionths of a second, it was the world's first pulse of laser light.

His device was so small and looked so simple that the laboratory where Maiman worked, Hughes Research, insisted on distributing publicity photos of a larger, more complicated laser machine that actually wasn't working yet. Maiman's device had the same two components found in all lasers: a central medium that emits the light and a pump for adding the necessary energy to excite the electrons in that medium. In this case he encircled his ruby core with a strobe flash lamp that bathed the core in white light. The core itself was a prototype for all lasers. To amplify the stimulated light and produce even more photons, he coated the flat ends of the cylinders with silver so they would reflect the parallel rays of light back and forth through the core. One end was only partially reflective; it emitted the excess light in a pure, fierce beam. As with all lasers, the actual light given off was really just a safety valve, only about 1 percent of the light produced. If the end was fully silvered, the core would heat up as it built up more and more photons, and eventually explode.

## The Early Lasers

Soon after Maiman's triumph, scientists began coaxing laser light out of an amazing assortment of substances. Indeed, Maiman would say after his success that scientists were putting mirrors on crystals of anything to see if it would "lase"—produce laser light—and a surprising number of times the crystals would. In the next few years scientists would build lasers with cores made of substances as diverse as the noble gases neon and krypton, metal vapors, and liquids containing organic dyes. When the common gas carbon dioxide was "lased" soon after the release of *Goldfinger* in the mid-1960s, an infrared workhorse was created that could actually cut things the way it had been done in the movie.

Some lasers, such as the first semiconductor laser built in 1962, needed to be plunged in liquid nitrogen to be cooled so they wouldn't explode. Others burped out light pulses that lasted only microseconds. Scientists used a variety of techniques to pump their lasers, including chemical reactions, electrical discharges via diodes on each end of the medium, and flashes of light including light from another laser. Argon lasers emitted beams of stunning greenish light, while most lasers shot out light either too energetic (ultraviolet light) or long in wavelength (infrared and microwave) for the human eye to detect. Yet despite their variety, all these lasers shared many characteristics.

The light they produced was often many times more brilliant than that from a comparable area of the surface of the sun, but was usually not of great wattage, or power—typically far less than that produced by a flashlight. The shorter the wavelength, the more energetic the light, but also the more energy needed to pump up the atoms of the medium: X-ray lasers, for example, would require two watts of power per atom for the needed population inversion to be achieved. With all lasers, what came out was always coherent and therefore useful in ways no other form of radiation could be. The entire power of a laser beam could be focused to a point well under a micron (1 millionth of a meter), gathering all the punch of the laser into an area smaller than a human cell. Also, because of its coherence, a beam of laser light would not spread out much over a long distance. Though laser light is not absolutely coherent forever and eventually diverges, by human standards it is of an unequaled purity and consistency. In 1969, for example, a 1-mm-wide beam from a ruby laser was bounced off reflectors placed on the moon. After its nearly half-million-mile round trip, the beam had widened to only a kilometer and a half.

## BRIGHT SPOTS: A FIELD GUIDE TO IMPORTANT LASERS

Masers preceded lasers because their longer microwave emissions required less energy to produce and hence they were easier to make. In early masers, wire screens served as the "mirrors" between which the photons bounced back and forth. The wavelength of a maser is typically one-hundredth of a meter.

With wavelengths between a millionth and a thousandth of a meter, infrared lasers produce photons that have enough energy to carry plenty of punch. The carbon dioxide laser, with a wavelength of about one-hundred-thousandth of a meter, is considered the industrial workhorse laser, and is important in surgery because its beam heats water.

Shorter in wavelength, the Nd:YAG laser (whose crystal, a member of the garnet family, consists of neodymium glass, a compound of the rare earth element yttrium and aluminum) emits light just outside the visible spectrum with a wavelength of about a millionth of a meter. These lasers drill holes for industry and are also important in eye surgery. They have an advantage over their red-light cousins, the ruby lasers (wavelength: about seven ten-millionths of a meter), because the YAG crystal cools relatively easily and thus can fire continuously.

Many gas-medium lasers produce light in the visible spectrum. These include the very common helium neon laser (wavelength about six ten-millionths of a meter), the krypton laser (also about six ten-millionths), and the argon laser (about five ten-millionths).

Other important visible-light lasers include dye lasers, based on a liquid medium of an organic dye. By changing the composition of the dye, the emerging wavelengths can be changed or "tuned." Typically the light they produce will be below one millionth of a meter.

Emitting ultraviolet photons even shorter in wavelength but more energetic, excimer lasers use molecules created when a rare gas such as argon or krypton is combined with another gas, typically chlorine, in an electric current. Excimer lasers have numerous medical applications, including sealing leaking blood vessels and blasting clear clogged arteries. The wavelength of an excimer laser is typically one ten-millionth of a meter. Excimer lasers do not heat tissue as other lasers do. Instead, the beam breaks chemical bonds and allows cells to be washed away without damaging surrounding tissue.

X-ray lasers have been constructed in which each photon has a hundred times the energy of the photon produced in a visible-light laser. First a giant optical laser, delivering 10 trillion watts of power in a billionth of a second, vaporizes selenium atoms and strips them of their outer electrons. Then the remaining inner electrons "lase," emitting X rays about a hundred millionth of a meter in length. A future practical use of this laser may be to create holograms of structures too small to record with visible light, such as individual molecules. Free-electron lasers, in which the electrons are pumped to lasing capacity by an array of magnets, were first developed in 1977. They can be tuned, in theory, to any frequency of the spectrum, from infrared to X rays. Because they are so versatile, they have become a popular research tool. Their capacity to generate laser beams at very high frequencies enables research into esoteric areas of quantum physics, where events are measured in trillionths of a second.

Lasers had other restraints. They didn't use energy efficiently—converting only about 2 percent of the energy added to a system into light (not much worse than a typical incandescent lightbulb, but a poorer ratio than the 10 percent of a fluorescent light). And despite their much-publicized ability to punch through razor blades, they really weren't very good at cutting through things because their focused beams punched out small, conical holes. Early industrial uses seemed a far cry from felling trees. They excelled in soft mediums, where they were used to punch pinholes in baby-bottle nipples and aerosol valves. Another major industrial use became pricking microscopic holes into cigarette paper for airflow control.

## Lasers as Nature's Straight Edge

Though lasers may have come up short as new versions of chain saws and bowie knives, they have found other niches. Soon after their invention, they proved unparalleled as tools for measuring and mapping. Small, easy-to-use helium neon lasers were perfectly adequate for this work. Within a decade of their invention, lasers became vital supplemental tools to the carpenter's chalk line and the surveyor's transit.

Because the beam from a laser goes straight and remains narrow, it can be used to plot near-perfect lines and, with prisms, angles. Lasers also are used to mark level planes, using prisms to rotate their beams onto surfaces. By placing a laser on one foundation post, a builder can construct other foundation posts that go up exactly to the same height. Similar alignment lasers can make sure floors are perfectly flat, or that a bulldozer cuts along exactly the right path.

Surveyors use lasers not only to create the straight lines important to their craft, but also to measure distances. A burst of laser light can be aimed at a reflective target as far as forty miles away. A computer measures how long it takes the light to return, and calculates the distance. Using this technique, mapping jobs in the Grand Canyon that would formerly have taken a hundred people a year to complete have been finished in three days by only two workers.

On a grander scale, these simple tools have also been used to measure movement of the Earth's tectonic plates. In one project, scientists bounced laser light off the orbiting Laser Geodynamic Satellite, to calculate the exact position of those plates. With repeated measurements, small movements of the Earth's surface could be detected.

Lasers bring similar precision to the microscopic level. Laser-based interferometers can measure differences in distances as small as a thousandth of a millimeter, which is a thousand nanometers. Here one laser beam is split and then rejoined later at a point called the interference point. When the distances the two beams have traveled are the same, the light will rejoin in phase. But as the distance is changed ever so slightly and the two waves arrive more and more out of phase, the interference point will dim, then brighten again as the waves realign. This laser is used to ensure quality control in the manufacture of precision machine tools.

Usually, helium neon lasers are the workhorse light source for these applications. Their vivid red light is familiar to us because they also are used in one of the ubiquitous triumphs of laser technology, the supermarket scanner.

## The Quantum Supermarket

Scanners date back to the mid-1970s, when the United States Supermarket Institute agreed on a labeling standard for all products called the Universal Product Code, which assigns two six-digit numbers to a wide variety of consumer products. The first digit identifies what the product is, the second its specific size. Now-familiar bar-code representations of these numbers are printed on packaging, and these are machine-readable without resorting to special inks or materials.

In the actual scanner, a beam of visible laser light is directed into a scanning pattern just outside its housing. An optical sensor inside the scanner waits for an item to be passed over and the laser light to be reflected back onto it. This light will blink on and off as the bar code passes over the beam of the laser. A chip in the scanner waits for the sensor to send a series of on/off signals that are a valid sequence for a bar code. It beeps its affirmation of a reading, and sends the corresponding number to the cash register.

## Lasers and Digital Information

Using a sensor to detect whether a beam of light is bounced back is a simple way to record information. Either the light gets reflected or it doesn't. What is stunning is just how much information can be stored this way per unit area, and how quickly it can be read, thanks to a laser beam's ability to get very small.

The development of laser disks, compact disks, and digital video disks (DVDs) awaited the discovery of a cheap, dependable laser source of safe, low power. Semiconductor lasers went back to the early 1960s, but it wasn't until 1975 that these became reliable and easy to make, and didn't require cooling in a nitrogen bath.

A semiconductor laser is tiny—not more than a millimeter in any dimension. Silicon does not emit light, so semiconductor lasers are made of compounds of gallium arsenide. The laser is a sandwich of positively charged and negatively charged material. Between is a small channel or *quantum well*. When a positive charge is placed on the positive layer and a negative charge on the negative layer, the electrons and positive particles dive into the well and then recombine, releasing light. (This is also all there is to a light-emitting diode, or LED.) With a laser, the sides of the channel are cut in such a way that the released light is reflected back and forth, stimulating the production of laser light.

The laser light that is the heart of a typical CD player is not visible. Its wavelength is about 820 nanometers or billionths of a meter, just beyond visibility in the infrared range. This beam shines on the spinning CD, where it is either reflected back onto a sensor or diffused by hitting a "pothole"—a small pit carved into the plastic. When light gets reflected back, the sensor sends out a one; if it isn't, the sensor sends out a zero. As the laser beam tracks the disk, the sensor gives off a flow of digital information.

A pit registering a zero in a traditional CD player can be as small as a micron—a millionth of a meter—or about one-hundredth the width of a human hair. Because the total length of the path of these pits and smooth stretches on a CD is about 5 kilometers, the number of bits of information that can be stored works out to 0.68 gigabytes or 680 million bytes of information.

If pits can be made smaller, a disk can hold even more information. But what limits this size, in turn, is how narrowly one can focus the reading laser beam. Newer DVD players have lasers that produce visible light of a shorter wavelength, which can be focused into beams that read pits in disks half the size of the traditional 1-micron limit. Because pits can be half their former width as well as half the length, the track of information also becomes narrower, and twice the length may be fitted onto the same size disk. One disk can also be designed to hold even more information by stacking two tracks on top of each other and placing tracks on both sides of a disk. Thus the 0.68-gigabyte capacity CD is redesigned into a 17-gigabyte DVD giant without increasing the size of the actual disk.

Engineers are confident that optical players will next rocket in capacity when they perfect machines based on lasers producing even more compact blue light, with a wavelength down around 460 nanometers, or slightly less than half a micron. The spot diameter of a blue laser is half that of an infrared laser. In addition to increasing the capacity of a compact disk, blue diode lasers will greatly increase the quality of laser printers. In 1997, scientists at the Xerox Palo Alto Research Center (PARC), where the first high-speed laser printer was invented more than twenty years ago, successfully generated a blue diode laser beam using a gallium nitride–based blue diode laser. The device is now being fine-tuned for commercial applications.

Semiconductor lasers are also an important element of optical fiber communications. Telephone conversations, video, Internet traffic, and other data are transmitted as streams of laser light pulses through glass filaments. Modulators that act as shutters block the light or let it through, turning the light pulses into the kind of lightning-fast digital information that is the basis of modern telecommunications. Because lasers can be tuned to produce light of a very specific wavelength, the capacity of fiber-optic networks has been increased greatly by the use of multiple wavelengths of light on a single fiber. Recently developed optical-networking systems can deliver more than 1 terabit (1 trillion bits) per second of information over a single strand of fiber (see chapter 3).

## Quantum Cascade Lasers

The quantum cascade laser is a speck of semiconductor material just a few millimeters long. But it is more powerful than a conventional semiconductor laser, and much more versatile. It can be tailored to emit light at a selected frequency within the infrared regions of the electromagnetic spectrum. This range of the spectrum is important for pollution monitoring and industrial process control for environmentally safe manufacturing because many hazardous and toxic chemicals absorb light rays at the wavelengths of these lasers. To monitor for the presence of even minute quantities of a specific chemical in the atmosphere, engineers can tune the laser to the appropriate wavelength, shine the beam through the air, and measure the amount of radiation absorbed.

The quantum cascade laser is aptly named, because it is designed to take advantage of the quantum behavior of electrons to build up a "waterfall" of photons from a relatively small input of electrons. This is accomplished by constructing multiple layers of semiconductor materials that conduct electrons down a series of energy levels. As it jumps between each energy level, the electron emits an infrared photon or light pulse. The emitted photons are reflected back and forth between built-in mirrors, stimulating other quantum jumps and the emission of other photons. This amplification process enables high output power.

The quantum cascade laser can be tailored to emit light at a specific wavelength set at nearly any point over a very wide range of the infrared spectrum. This is done in the manufacturing process, by varying the thickness of the layers of semiconductor materials used to construct the laser. The thickness of these layers ranges from between just three to forty atoms. They are made by a process called *molecular-beam epitaxy* (MBE), which facilitates the construction of complicated semiconductor devices one atomic layer at a time (see chapter 8). The first quantum cascade laser was invented in 1994. But its utility was limited because early quantum cascade lasers operated at temperatures lower than about −300 degrees Fahrenheit (−180 Celsius). The first room-temperature and high-temperature quantum cascade laser was developed in 1996 at Bell Laboratories, part of Lucent Technologies. With this breakthrough, practical devices based on quantum cascade lasers are likely to be developed.

## Holograms: How Lasers Help Deliver the Complete Picture

Holograms have become part of our everyday baggage: we often carry a small hologram on a credit card, and wrap party gifts in hologrammed paper. Few are aware of the sophisticated optics and laser technology that go into their creation. A hologram works analogously to the way the human visual system perceives objects in our three-dimensional world. In the 3-D world we see, light waves strike countless points on an object, which bounce back and disperse these waves so they cross, cancel, and add to each other in an enormously complex, choppy pattern. Holograms try to capture a slice of this complex interaction of light coming off an object—a piece of what is called its wavefront. By contrast, a two-dimensional photograph captures the intensity of a limited selection of light wavelengths. Holograms produce three-dimensional images by capturing certain wavelengths of light coming off an object when the light striking the object is coherent laser light.

The standard way to make holograms is to split a beam of laser light. One beam, called the *object beam*, is passed through lenses and projected onto an object. Nearby is a piece of very high resolution film, which is positioned to become the "window" at which the light passing through will be "captured." The way this is done is to have the second beam, called the *reference beam*, also shine on the film, coming in at more or less right angles to the light coming off the object. The system of stripes, bright patches, and dark spots created when these two light sources interact is then burned into the film.

To re-create the object as a hologram, the reference beam is shone on the film that has captured this interference pattern. Some of this light passes right through. But some bounces off the tiny stripes and patches, which diffract it at an angle that matches the light that was striking the film from the object. In fact, this diffracted light will exactly re-create that slice of the wavefront that was coming from the object at the moment the film was exposed.

Different types of hologram are distinguished from each other by the way diffraction is used to re-create the image. When light passes through an exposed hologram, it is called *transmissive*, while if the diffracted light gets bounced back, it is *reflective*. White-light holograms record wavefronts from three primary-color lasers and "play them back" at the same time. Rainbow holograms will reflect images at different wavelengths of normal white light, but in different places—hence the image will change color like a rainbow if the hologram is moved slightly. Holographic movies are attempts to capture a series of wavefront slices and then play them back to capture movement. Holograms can now store a few thousand images on top of each other—in a film or photosensitive crystal—and move as the viewer moves or a slightly different hologram is projected. In the future, they may become the medium of choice for a fast, high-capacity storage system for computer information (see chapter 1). And X-ray holograms may soon freeze three-dimensional images of objects too small to be viewed by optical microscopes.

Currently their main industrial application has been holographic interferometry, in which tiny changes to an object can be observed by comparing two holograms of it. By observing holograms of something before and during the time it is subjected to some force or stress, engineers can see how much contortion and disfigurement it is going through. More startling, using complex formulas, computers are able to calculate the diffraction patterns needed to produce a hologram—and hence design holograms that project images of objects that were never "photographed." Thus they can create artificial but totally compelling 3-D vistas that never really existed.

## Scalpels of Light: Lasers in Medicine

Lasers became important in information storage because their beams were bright and could be precisely focused. The implications of this exactitude have also not been lost to those building equipment in the important human realm of curing and preventing diseases.

With the invention of the *Goldfinger*-era carbon dioxide laser, which focuses light rays into a very concentrated beam that is ideal for precise cutting or vaporizing tissue, there were predictions that scientists were on the verge of coming up with "knives of light" that would replace the steel scalpel, but scalpels are small, light, relatively cheap, and cut flesh well. Lasers instead have found other uses for which they are considered indispensable.

Doctors have learned to exploit not just the precision of the laser beam but also its uniformity. Because lasers emit intense beams of light of one specific wavelength, the energy in a laser will only be transferred to certain pigments and colors while other materials will be completely un-

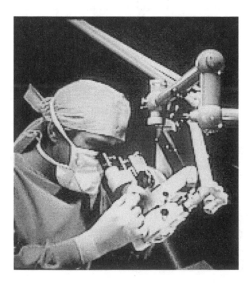

**Figure 4.2** Microsurgery using a carbon dioxide laser is practiced in a variety of surgical specialties, including gynecology and neurosurgery. The device is attached to the operating microscope.

affected. With a laser, you can select which tissue it is you want to attack, while, with proper technique, leaving other tissue unscathed. Laser surgeons contend that laser surgery is safer than traditional surgery involving knives or scissors. Because there is no incision, there is less risk of infection and bleeding. In addition, most laser surgeries are performed on an outpatient basis with little postoperative pain and recovery time.

Medical lasers first found practical uses in dermatology and ophthalmology. Pioneers in laser ophthalmology had a precedent to their work; in the 1940s the German scientist Gerd Meyer-Schwickerath used sunlight to treat detached retinas and tumors in the eye. Three years after its invention, the laser was being used to treat retinal diseases. The light from a laser would pass through clear eye structures without affecting them in any way, only to excite the darker retina. Ophthalmologists frequently make use of the Nd:YAG laser (see page 93), which is best for its ability to coagulate blood and seal blood vessels. To treat certain kinds of glaucoma, ophthalmologists use the Nd:YAG laser to put holes in the iris, allowing the fluid to drain out.

The ability to heat up specifically colored pigments was a property also exploited in using argon lasers to remove red birthmarks, also called portwine stains. The beams of light pass harmlessly through the skin to excite the thick net of extra blood vessels underneath, and heat them to the point they are obliterated. Unfortunately, once this heat is created, it's hard to control and can damage surrounding skin, discoloring it and scarring it. The laser treatment was refined in the 1980s, when doctors realized that if pulses of

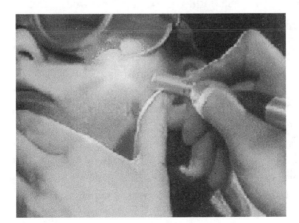

**Figure 4.3** Argon lasers treat port wine stains and other disfiguring skin lesions. Tattoos that have lost their allure can be removed by argon, ruby, and carbon dioxide lasers.

laser light were kept short, under a thousandth of a second, target areas would receive the same doses of energy but have the chance to dissipate their heat less destructively. This technique of using pulses to heat target tissue is now also used in tattoo removal, in which the pigments of injected dyes disintegrate from the energy they absorb from laser light delivered in pulses.

The selectivity of what a laser heats is also being tested to fight cancer using a new approach called photodynamic or photoradiation therapy (see chapter 6). This technique uses chemicals that are activated by a specific wavelength of light to kill the cells in a tumor. Researchers use compounds that are much more readily absorbed by some tumors than they are by normal, healthy cells. Laser light is then used to activate the chemical in the tumor. Researchers have even used laser light transmitted via endoscopic instruments to reach tumors that are buried deep inside the body.

Laser light carried through optical fiber tubing is also being used to seal blood vessels and shrink blisters deep in the lungs. This helps in the treatment of severe emphysema and other pulmonary diseases. Similarly, thin optical fibers now deliver pulses of excimer laser light to the walls of blood vessels serving the heart, where it can obliterate blockages and restore circulation. Thousands of these "laser angioplasties" have been performed.

Surgeons do sometimes welcome a laser's ability to heat things very fast. The most general laser scalpel to be invented is one that uses the infrared light generated by carbon dioxide. These lasers work like tiny microwave ovens as water molecules absorb their wavelength readily and quickly turn into steam. This means that as it "slices," a carbon dioxide laser seals up and cauterizes blood vessels, reducing bleeding. If a surgeon needs to remove just the damaged portion of a liver, for example, she will likely use a carbon dioxide laser because the heat it produces will close the capillaries it cuts through. Newly developed hand-held carbon dioxide lasers have recently been put to use to perform tonsillectomies. The use of a laser to remove enlarged tonsils is considerably less painful than traditional methods, because it removes less tissue. Lasers have also been used to remove excess flesh in the palate and the back of the throat to restore a measure of peace to the nighttime companions of loud snorers.

Lasers are doing more and more delicate work in the part of the body where it all began—the eye. Argon lasers can seal bleeding retinal blood vessels, a serious condition in diabetic retinopathy that can lead to blindness. They are also used to "weld" detached retinas back in place. YAG crystal lasers are used to remove what are called secondary cataracts, which often form behind an artificial lens implant. Shock waves created as the faulty membrane absorbs the infrared light will tear it apart.

One of the most exotic and dramatic procedures using a laser will likely also become the most common: correcting nearsighted or farsighted vision by "sculpting" the cornea, which is called photorefractive keratectomy. First the skin covering the cornea is peeled back. Then short, high-energy bursts from an ultraviolet excimer laser "sculpt" the lens by shaving off a layer of the cornea. A temporary soft contact lens is placed on the cornea to act as a bandage, and the patient is monitored over the next few days to make sure the cornea is growing back in the correct shape. This growth can be adjusted with medication. Photorefractive keratectomy, also known as PRK, holds the promise of allowing many nearsighted or farsighted people to throw their glasses or contact lenses away. The long-term efficacy of the procedure, however, has not yet been established.

## Death Beams and Ray Guns: The Military Uses of the Laser

In *The War of the Worlds*, written in 1898, H. G. Wells gave his attacking Martians a big edge in military hardware. They leveled whole cities quickly with a "flaming death," an "invisible, inevitable sword of heat." Military strategists have had their eye on killer rays ever since.

Lasers that can shoot down an airplane or pierce the armor of a tank have indeed been developed, but they are bulky and limited in the number of times they can fire. The problem is that the power in a beam ultimately comes from whatever pumps the lasing material, and to get a high amount of power you need a big pump. For killer-strength beams, weapon planners have come up with chemical lasers primed with chemical explosives. But these are an inefficient use of a chemical reaction, as most of the energy will not be transferred into the laser beam. And the combustible components of a chemical laser, typically fluorine and hydrogen, are easy to set off accidentally. Billions have been spent studying the idea of building a space-based laser or particle-beam system that can shoot down intercontinental missiles. In one recent incarnation, this idea would place twenty-five "killer satellites" in orbit, which would monitor the entire world for enemy missiles. The price tag for the system was pegged at at least $30 billion.

Weapons engineers believe such beams are entirely possible, but computer experts are dubious that any such system could be made reliable. Needing to detect and react in minutes to enemy launches, the computer programs running the weapons would have to be fantastically complex and then would require billions of instructions. And testing the system thoroughly would be impossible without staging practice missile attacks against the United States.

Though so far impractical as a medium for ray guns and death beams, lasers have other military uses that are indispensable. As the public learned during the Persian Gulf War in 1991, "smart" bombs and missiles lock onto targets with lasers, which guide the delivery of these weapons. Laser gunsights can help soldiers aim their weapons with deadly accuracy. These devices are becoming more common as additions to the policeman's standard service revolver, and have even been used on an experimental basis by cops patrolling the New York City subways. Other military uses for the laser include "eye popper" beams to blind orbiting spy satellites by directing the laser beam at the satellite's sensing devices, and as water-penetrating "flashes" for special cameras designed to take pictures of objects underwater. Designed to study sunken Soviet submarines, these laser cameras have had civilian uses such as analyzing underwater airplane wrecks. One was used extensively to explore the wreckage of TWA flight 800, which plunged into the Atlantic off Long Island in 1996.

## Lasers in the Future: Going Microscopic

More experimental laser uses involve manipulating individual cells. In fertility research, for example, lasers have drilled holes into rabbit eggs to make them easier for sperm to penetrate, and laser "tweezers" have been used also to pick up individual sperm cells and introduce them to the egg. Tiny two-to-four-cell human embryos, products of in vitro fertilization, have their tough outer layers shaved away with laser light to improve their chances of attaching to the uterus when implanted back in the womb. Scientists have trapped single human sperm cells, then lowered the intensity of the beam that holds them until the cell can swim away, thus obtaining an exact reading of its power as a swimmer. In other research, scientists used laser beams to trap the vesicles of a sea slug in order to study the cells' chemical contents. Vesicles are tiny cells that carry chemical messages between neurons. These procedures are part of the growing use of lasers in the microscopic realm. In the future, microscopic laser "scissors" may be used to cut holes in cell nuclei, for example, to insert foreign genes into the DNA of the cell. Laser "tweezers" may be used to stretch out single chromosomes so they in turn may be snipped into shorter strands.

## Creating "Optical Molasses"

Physicists exploring the esoteric realm of quantum mechanics have used lasers as tweezers to immobilize individual atoms. At normal temperatures, atoms vibrate and shoot around at very high speeds. Air molecules, for example, typically travel at speeds faster than the speed of sound.

With lasers, physicists have managed to trap and immobilize small groups of atoms. They call this procedure "optical molasses."

Laser tweezers make use of a characteristic of light we usually don't notice because the light we encounter is relatively diffuse and because we're big. Light exerts a force. People are too big, and the light around them too weak, for this force to be significant. At atomic levels, however, laser light can immobilize a tiny object in the same way that a mounting pin holds down a butterfly.

Using "optical molasses," scientists have slowed atoms down to a state that approaches immobility, thereby lowering the temperature of those atoms to only a few millionths of a degree above absolute zero, the temperature at which atomic movement stops. At such low temperatures, with their momentum disappearing, atoms start acting like long waves and have ill-defined locations. This soup of fuzzy stuff, half waves and half particles, is called a Bose-Einstein condensation. Its existence was long predicted by quantum mechanics, which held that as a particle's speed and hence its momentum dropped toward zero, two things would happen—its exact location would become harder and harder to determine, and it would become a longer and longer wave.

Lasers allowed physicists to create Bose-Einstein condensation in the early 1990s. Now this substance has been used in laboratories to create a new type of laser, the atom laser. Using lasers and magnetic traps, scientists have cooled sodium atoms into condensation. Then, because this condensation acts like a wave, they have managed to shoot "drops" of these waves in brief pulses, each holding the equivalent of about a million atoms. No practical application is near for the atom laser, though scientists say similar ones may be able to act like a matter "cake decorator"—squirting a few atoms at a time to build structures atom by atom. A future application could be to create electrical circuits with dimensions that are measured in atoms.

Lasers are already being used to prod atoms to make them combine in desired ways. Traditionally, chemists create complex chemical compounds using heat, pressure, or an enzyme or catalyst. But at the atomic level, an atom's behavior is never completely predictable. Often atoms have more than one choice for the way they interact and the compounds they form. By shining the right wavelengths of light on atoms, patterns can be formed that reduce the chances of one "choice" being created and therefore raise the likelihood of the other path being taken. Using lasers, scientists have been able to raise the chances that two atoms will form one compound rather than another from 50 percent to as high as 90 percent. This technique may greatly reduce the cost of producing complex chemicals such as pharmaceuticals, where there are many wrong paths that molecules can take when combining to form a desired substance. (See also chapters 8 and 9.)

# A GLOSSARY OF LIGHTWAVES AND LASER TECHNOLOGY

**coherence**  A quality attributed to electromagnetic radiation that emerges at one wavelength with wave crests and troughs aligned. Laser light is coherent, while the light emitted by a typical incandescent lightbulb, for example, is incoherent.

**hertz**  A measure of the number of cycles per second of an electromagnetic wave. Light is typically measured in millions of cycles per second, or in megahertz (mHz).

**hologram**  A method of capturing the entire wavefront of an object and then reproducing this wavefront, creating an image that will appear to be three-dimensional. At present, holograms are limited in the amount of color they can reproduce.

**maser**  The first type of laser: a maser is a stimulated emission device that emits microwave radiation.

**monochromatic light**  Light of a single wavelength and hence only one color. The coherent light of a laser is monochromatic.

**optical molasses**  A trap for molecular and atomic particles created with the momentum of light.

**optics**  The branch of physics that deals with light and vision, optics is mostly based on the wavelike properties of light.

**photon**  The basic unit of electromagnetic radiation.

**photorefractive keratectomy**  Correcting nearsightedness or farsightedness by using a laser to correct corneal imperfections by "sculpting" a thin layer of cells.

**population inversion**  The description of what happens to a group of atoms when they are ready to produce laser light; almost all the atoms have electrons in a higher energy state.

**pumping**  The process of adding energy to the core of a laser so that its atoms will invert and produce coherent radiation.

**spontaneous emission**  The electromagnetic radiation generated when an electron drops from a higher energy state to a lower energy state.

**stimulated emission**  The electromagnetic radiation generated when an electron in a higher energy state is struck by a photon from a similarly excited atom. The source of laser light.

**wave properties**  The properties that electromagnetic waves exhibit are reflection (bouncing off substances), refraction (having their direction changed by passing through a medium), diffraction (having their direction deflected by part of the wave encountering an obstruction), and interference (having their peaks and troughs added to and subtracted from as they cross each other).

# 5

# All in the Genes: DNA Becomes an Industry

Billions of years before the first computer engineer created transistors based on the ones and zeros of open and closed electrical circuits, nature devised a code that is the ultimate arbiter of life on Earth in all of its immense diversity and complexity. The deciphering of this code is arguably one of the most important discoveries of the twentieth century. It has revealed that the life processes of every living organism, from the bacterium to the blue whale, are governed by the same four chemical compounds, called nucleotides, joined together in varying lengths and sequences along a phosphate spine to form deoxyribonucleic acid, or DNA. This discovery, that all living things have so much biochemistry in common, is what has enabled the burgeoning field of biotechnology in general and genetic engineering in particular to develop.

In the early nineteenth century, well before the genetic code was deciphered, biochemists began analyzing and synthesizing the materials produced by the enormously productive chemical factory that is a living cell or microorganism. They used an early incarnation of biotechnology—the manipulation of microorganisms and enzymes to produce commercially useful products—in enterprises that ranged from brewing beer to making drugs. Now, in the decades since the Nobel Prize–winning biochemists James Watson and Francis Crick definitively established the structure of DNA in 1953, researchers are probing and manipulating the genes controlling this factory, and their work holds profound implications for human life and commerce. Indeed, many observers believe that we are at the beginning of a "genetic revolution" that will have as much impact on society as the invention of the steam engine and the microchip.

Working in such fields as materials science, agriculture, medicine, and pharmaceuticals, genetics researchers have been able not only to decipher the genetic code, but to replicate and manipulate it. They have discovered that it is possible to synthesize stretches of DNA and to snip out pieces of it at a precise point from any genome. They have developed methods of deciphering the DNA code of increasingly complex living organisms and of using elements of DNA to create diagnostic tools and unique DNA "fingerprints" for every living creature. They have also devised ways to insert "foreign" DNA into the genome (or the set of DNA) of another organism and thus manipulate the products of living organisms.

A broad array of DNA-related activities is the focus of intense scientific study and commercial activity. Worldwide, new and established companies and research institutes, whose senior managements often include Nobel Prize winners, are assiduously studying and manipulating DNA to develop diagnostic tools, produce drugs and new materials, fight such previously intractable diseases as cancer, obesity, and old age, and create more productive and disease-resistant plants and animals. On an almost daily basis, the scientific and popular media report the discovery of new genes and the development of novel genetic engineering methods and products. For the general public, these developments are both exciting and frightening, as the secrets of life itself, which some consider sacred, are revealed and manipulated. It is an area so fraught with ethical dilemmas that one major development in the field, the reported cloning of an adult sheep by a Scottish researcher in 1997, quickly resulted in a call by the President of the United States for a commission to review the "troubling" implications of the procedure, and elicited proposals for legislation to limit it.

An important element of genetics research is the prospect for huge financial payoffs to the companies that can exploit the potential of biotechnology to develop effective cancer therapies or super pest-resistant food crops. Indeed, the growing store of information about the DNA sequences in specific genes or of a particular organism has become a commodity in itself. Private companies, universities, and research institutes have patented potentially profitable genetic sequences as they are deciphered. A new discipline, bio-informatics, uses sophisticated computer database programs to help interpret the complex biological data generated by DNA research. Thus it is fitting that Arthur Kornberg, who received the Nobel Prize for Medicine in 1959 for synthesizing DNA, has described this long and infinitesimally thin molecule as "the golden helix."

## SPONTANEOUS GENERATION AND BEYOND: SOME EARLY MILESTONES IN THE SCIENCE OF GENETICS

Some of nature's most fundamental conundrums were solved relatively late in the history of mankind. Well into the nineteenth century, eminent scientists believed in "spontaneous generation," i.e., that life somehow arose from a mixture of dust and something called a "vital force." Explaining how offspring inherited the characteristics of their parents, alchemists in the Middle Ages believed that an egg contained a preformed small human being, a "homunculus" that was passed on through the generations.

Even in the fifth century B.C., the Greek philosopher and physician Hippocrates observed that traits and diseases had a hereditary component, but it was only in the seventeenth century that Robert Hook first described "cells," and a Dutch scientist, Antonie van Leeuwenhoek, whose hobby was constructing simple microscopes, observed microorganisms like bacteria. Louis Pasteur, renowned for his seminal work in disease theory and immunization, definitively put the belief in spontaneous generation to rest in 1864, with a historic series of experiments.

In 1859, Charles Darwin published his *Origin of Species,* presenting his theory of evolution by natural selection, and in 1865, Gregor Mendel, an Austrian monk, established many of the principles of genetic inheritance in his studies of garden peas. Mendel ascribed general principles of dominance and recessiveness to genes without actually knowing what they were. In 1869, Friedrich Miescher discovered the nucleic acid DNA in the nuclei of cells from pus, and ten years later a German scientist, Walther Flemming, discovered tiny, threadlike structures within the nucleus, made up of a material that he called chromatin. The stained material, now known to be DNA, complexed with protective proteins, revealed the details of mitosis (cell division) to Flemming and others.

The 1940s and 1950s saw many of the most important developments in genetics, when Oswald T. Avery, a medical microbiologist at Rockefeller University in New York, working with the bacterium *Pneumococcus pneumoniae,* determined conclusively that the means of heredity was DNA. In 1953, Francis Crick and James Watson, interpreting X-ray photographs of DNA, described its double helical structure and suggested that this structure allowed the DNA molecule to replicate in a very simple way. In 1956, Crick proposed that information in most organisms flows from the DNA to protein synthesis. He also suggested the mechanisms by which the protein-synthesizing instructions are carried out. During this period, researchers began to elucidate the role of ribonucleic acid (RNA), the intermediary molecule that carries out the instructions encoded by DNA within the cytoplasm (see page 113). In 1961, Crick, working with Sydney Brenner and other researchers, established how the genetic code translates into instructions for assembling proteins in the cell.

## Necklace of Life: The ABCs of DNA

Magnified millions of times, the DNA molecule would probably look like a long, double-stranded necklace made of just four different interlocking beads. The beads are bases that belong to two families called *purines* and *pyrimidines*. The purines in DNA are adenine and guanine (A and G), while the pyrimidines are cytosine and thymine (C and T). A close look at the necklace would reveal that a particular bead on one strand was always joined to a specific bead on the opposite string. It is almost as if the one bead is shaped in a way that fits exactly into its partner. These beads are complementary bases that always connect in the same way: adenine with thymine, and cytosine with guanine. Thus, for example, a strand of DNA that reads AAGTCCTAA will have a complementary strand that reads TTCAGGATT. This complementary bonding is called *base pairing*.

When a cell divides, the double strand of DNA "unzips," and then a complementary strand of DNA forms and links up with its partner. Then the cells split into two new cells, a process known as *mitosis*. This is the way that genetic information is passed on through generations of cell division. Many consider the concept of base pairing the most powerful in molecular microbiology because it explains how DNA is replicated and how DNA directs the synthesis of proteins in the cell. Base pairing is also the keystone of most genetic research and genetic engineering methods because scientists use their knowledge of base pairs to unlock the codes of the DNA they are studying and to duplicate known sequences of the molecule. If the sequence of bases in one strand is known, then the sequence of bases in the other can be inferred.

What is the information contained in DNA? Geneticists have made much progress in the long research journey that will reveal how DNA orchestrates the thousands of interdependent chemical reactions that take place in a living cell. There are still many gaps in their understanding. They do know that specific long sequences of bases in a strand of DNA contain the instructions that a cell uses to assemble a particular protein, in much the same way that the varying sequences of the twenty-six letters

**Figure 5.1** When DNA is replicated, the double-stranded DNA molecule unwinds, and the separated strands serve as a template for the assembly of a new strand of DNA. For example, a strand that consists of the bases adenine (A), guanine (G), thymine (T), and cytosine (C) will be complemented by T, C, A, and G. The complementary bases are linked to each other by specific types of weak hydrogen bonds. Adenine and thymine are connected by two hydrogen bonds, while guanine and cytosine are connected by three. The two resulting double helixes thus formed are identical to the parent.

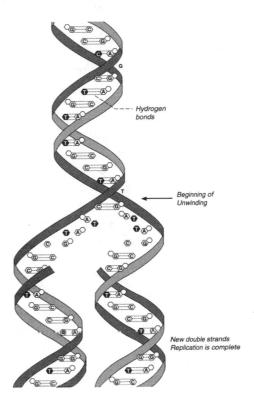

of the English alphabet form words and sentences. These long sequences of bases are called genes.

## The Universal Code of Life

Most genes direct the synthesis and functioning of proteins. These large, complex molecules are the constituents of many structural and functional elements of a living organism. Hemoglobin, the component of red blood cells that carries oxygen to different parts of the body, is a protein, as are the countless enzymes that digest our food and mediate the various biochemical pathways of our body functions. Antibodies that fight infections are proteins, as are hormones and the ingredients of our nails and hair. Even the substances that ensure the integrity of the DNA itself are proteins. Although every cell in our bodies contains the same set of DNA instructions, only a certain number of these genes are active in a particular type of cell. Genes that are active in a cell are "expressed" or "switched on."

How do our genes direct a cell to assemble a particular protein? The instructions are contained in the order of the nucleotide bases of a particular gene. This order of bases is said to "code" for a particular protein in the same way as a particular series of dots and dashes in Morse code translate into a specific letter or word. The DNA does not code directly for a whole protein, but for the sequence of amino acids that are linked together to form proteins. One of the most fascinating and fruitful discoveries of microbiology is that all the different proteins in every living organism on earth are constructed from a selection of just twenty amino acids. The universal code of DNA translates into a code of amino acids. In every living organism, from viruses to giraffes, the same three bases of DNA, called *triplets* or *codons*, code for each amino acid.

Marshall Nirenberg and Johann Matthaei were the first to crack the amino acid code. In 1961, while studying RNA (see below), the scientists deduced that a sequence of UUU—three uracils in a row—produced the amino acid phenylalanine. Inspired by the work of Nirenberg and Matthaei, Crick and Brenner established the basis for the genetic code of all proteins. In subsequent years, researchers continued to unravel the code by preparing nucleotides in various orders and detecting which amino acids the various sets of three bases produced. By 1967, researchers Har Gobin Khorana and Marshall Nirenberg finally completed the work. They were able to reveal that sixty-one varying triplet sequences (some amino acids have more than one triplet code) code for the twenty amino acids that make up all the proteins in our living world. Three additional triplet sequences, called *stop codons*, signal the end of a DNA sequence that codes for a particular protein. Another triplet sequence, which codes for the amino acid methionine, signals the start of the sequence. Most proteins are constructed from at least one hundred amino acids strung together and "folded" into shapes that determine their function, and some consist of thousands of amino acids. Of course, the varying combinations of these twenty amino acids can produce millions of different proteins. Once a gene is sequenced, knowledge of the amino acid code deduced from the sequence enables the researcher to figure out the protein it codes for. And vice versa: Knowledge of the structure of a certain protein allows a researcher to seek out its DNA, and possibly to study the neighboring DNA and its role in activating this particular gene.

Organisms that range from the bacterium *Escherichia coli* to plants and humans use the same DNA template and triplet code to translate the messages in their genes. This common code, which has endured for billions of years, is one of the strongest proofs that we have for the common ancestry of all life. And it is the commonality of the amino acid code that allows for genetic engineering: Theoretically any gene, whether human, plant, bacterial, or viral, can be inserted into the genome of another organism and, under the correct regulatory mechanisms, can direct the synthesis of the protein for which it codes. Mostly, in practice, the gene is inserted into a bacterium, yeast, or plant cell, which then proceeds to synthesize the protein. This is enormously useful if the protein that needs to be synthesized is important for medical or commercial reasons and can only be obtained in small—and expensive—quantities using other methods. Thus, one of the first feats of genetic engineering, successfully carried out in 1982, was to manipulate the DNA of *Escherichia coli*, a bacterium, to manufacture insulin, which before this had to be laboriously extracted from the pancreas of a pig or cow.

## Molecular Messengers: The Role of RNA

The genetic information of plants and animals is stored in the cell nucleus, while proteins are synthesized in ribosomes, which are minute particles scattered in the thousands throughout the cell cytoplasm. If DNA never moves out of the nucleus, how are the instructions encoded in genes transmitted to the ribosomes? The instructions are conveyed from DNA to the ribosomes by an intermediary called ribonucleic acid, or RNA. RNA actually has almost the same chemical structure as DNA, the difference being that RNA does not use thymine. Instead, it uses a chemical called uracil (U), which forms a base pair with adenine, just as thymine does. Also, RNA is usually single-stranded.

If one thinks of the double-stranded helix of DNA as sets of interlocking beads that always pair with each other in the same way, one can picture RNA as a mobile strand of beads that is strung together in the nucleus when a gene is being expressed. The order of the beads is determined by one strand of the DNA necklace, called the *antisense strand*. The type of RNA that carries instructions from a gene to a ribosome where the protein is assembled is called *messenger RNA* (mRNA). In protein synthesis, an enzyme called *RNA polymerase* manufactures mRNA using a section of DNA as a template. The RNA polymerase proceeds along the gene, "unzipping" a small section of the DNA double helix of the expressed gene at a time. In the process, the enzyme then lines up complementary RNA nucleotides opposite one of the DNA strands. RNA polymerase then joins the nucleotides together to make a strand of mRNA. When the mRNA molecule is complete, it slides away from the double helix and heads for a ribosome.

How does the RNA polymerase enzyme select the DNA strand it will work with? The two ends of a single DNA strand are chemically different.

The strand that the nucleotides line up against is the template or antisense strand. The other is the coding or *sense strand*. The stretch of DNA that codes for an entire amino acid sequence is called an *open reading frame*. It takes about a minute for RNA to make a copy strand of the 1,350 bases in the gene for hemoglobin. The process is called transcription, and any interference with it can be fatal. Alpha-amanitin, the toxin in the death-cap mushroom, for example, does its deed by forming a tight bond with RNA polymerase, thus preventing it from carrying out its task. Genes do not generally act alone, but are usually dependent on other nearby regulatory sequences of DNA and proteins that promote their expression. When genetic engineers transfer genes from one species to another, they need to make sure that the sequences that "switch on" gene expression are also present. At this stage of genome research, the DNA sequences or other biochemical triggers that "switch on" the expression of many genes are not yet known.

Studies of DNA and RNA have revealed that stretches of many animal and plant genes do not code for any amino acid, or regulate transcription in any obvious way. There are long strings of apparently random series of bases between genes and within the genes themselves. As the mRNA is strung together, the noncoding DNA is clipped out. The "active" sections of the gene are called *exons*, while the intervening stretches are called *introns*. Some scientists refer to introns as "junk." But although their function is unknown at present, some geneticists believe they may in fact help regulate gene activity. Whatever their function, introns already play an important role in gene research, where they are widely used for genetic diagnostics and identification.

---

### ENZYMES: THE MATCHMAKERS OF NATURE

Enzymes are natural catalysts that enable most of the chemical activity of a cell (but are themselves unchanged in the process). Most enzymes (whose name always end in *-ase*) catalyze a specific reaction, including digesting food, making more DNA, joining small molecules, or cutting big molecules into smaller ones. They act like natural matchmakers, providing a site for various chemical reactions in a cell, and then slipping away to facilitate more reactions. There are thousands of different enzymes for all of the thousands of different reactions in a cell. In the extraordinarily intricate chemical pathways of the cell, enzymes both build DNA and are encoded for by specific genes. Many are so crucial to cell functioning that their absence (caused by a cell toxin or genetic mutation) can result in death.

## Restriction Enzymes, Base Pairs, and Sticky Ends: The Genetic Engineer's Toolbox

The rapid advance of genetic research and engineering technology is evident in the almost daily news of gene discoveries and engineering breakthroughs. One may wonder how exactly a genetic engineer can accomplish such feats on a molecule that is so extraordinarily long. The DNA in a single human cell would, if unwound, span almost six feet. The nucleus of a human sperm or egg cell contains 3 billion base pairs of DNA; non-sex cells, like those of the muscles, contain twice that number. Daniel H. Farkas, a molecular pathologist, has calculated that all the DNA in the average adult would, if strung together, stretch between the sun and the Earth more than 450 times!

To the untutored eye, a stretch of letter codes representing the nucleotide bases in a cell's DNA would resemble a large dictionary filled with page after page of A's, G's, C's, and T's, or a pyramid covered in Egyptian hieroglyphics in which just four pictograms are used in the inscription. Moreover, there are no spaces between the code letters, which could act as a deciphering aid. But, just as hieroglyphics have revealed their secrets, microbiologists have learned to read the messages contained in nucleotide sequences. Indeed, using knowledge based on a few basic principles first discovered and developed by "giants" like Francis Crick and Sidney Brenner, molecular biologists are rapidly developing an extensive set of tools for sequencing and manipulating DNA. Many of the processes in the laboratory are likened to following complicated culinary recipes. The youth and rapid advance of genetic research is demonstrated by the fact that one of the first "manuals" of genetic engineering—a relatively slim volume—was published in 1982. The latest edition of this publication consists of three volumes, each about eight hundred pages long. Other publishers mail a folder of updated protocols to subscribers monthly. In 1982, a professor of genetics recalls, a student could earn a doctorate by sequencing one thousand bases. "Now, forget it," she says. "I expect a Ph.D. student to sequence two thousand bases in two days without using automated methods."

The principles of genetic diagnosis and manipulation are based on fundamental knowledge of the structure and chemistry of the DNA molecule. Probably the first and most important concept is base pairing or complementarity: adenine will always join with thymine, and guanine will always join with cytosine. Base pairing is the foundation of most genetic research and genetic engineering methods because scientists use their knowledge of base pairs to unlock the codes of the DNA they are studying and to duplicate known sequences of DNA.

One method that exploits the principle of base pairing to identify a sequence of bases is to use short DNA probes that have been artificially synthesized so that their particular sequence is known. These strings of synthetic DNA are known as *oligonucleotides,* or "oligos" for short. The known sequence is tagged in some way, by color, radioactivity, or fluorescence. The DNA to be identified, known as the "target" DNA, is denatured, or separated from its complementary strand, by heating or with a strong alkali. It is then placed onto a gel with the probes, which seek out and bond with their complementary sequence. When a DNA probe joins with a single string of target DNA, it is said to hybridize. The hybridized DNA is then analyzed using light or color-detecting equipment. If the probe is radioactive, the DNA is carefully spread onto a piece of special filter paper. The paper is placed in contact with X-ray film. Each probe and its bonded gene show up as a dark spot. The state of the art in sequencing is such that a genetics researcher can compare prices for oligo probes among various suppliers, specify the oligo sequence needed, and receive it by courier the next day!

The principles of base pairing are also used in genetic engineering to splice pieces of DNA—be they genes or just a stretch of base pairs—to build what is called recombinant DNA. This is one of the methods used to introduce the genes from one species into the cells of a host organism of a different species. First the stretch of DNA is snipped out of its original molecule. This is accomplished by some of the most important compounds in the genetic engineer's toolbox: a range of enzymes called *restriction enzymes.* The first restriction enzyme was discovered by chance in 1970, and hundreds are now commercially available. These enzymes, used by bacteria to chop up the DNA of invading viruses, have become a vital tool in genetic manipulation. Each restriction enzyme, named after its bacterium of origin, cuts DNA at a specific sequence. Using all the restriction enzymes available, one can cut and paste any stretch of DNA into pieces that are up to a few thousand base pairs long. When pieces of DNA are cut by restriction enzymes, the resulting double-stranded DNA is often left with a few single bases at each end. In some cases these are "sticky ends" that will splice with a new strand of DNA as long as it has the appropriate complementary bases at its "sticky ends." To complete the splicing, an enzyme called DNA ligase, whose normal job in the cell is to keep DNA in good repair, is used.

When researchers probe for gene expression—or signs that a particular gene is active—they use base-pairing methods to identify the messenger RNA that ferries the instructions spelled out by DNA to the ribosomes. Only "active" genes synthesize messenger RNA. This is particularly useful because messenger RNA "clips" or edits out the noncoding DNA that is interspersed in a gene, so researchers do not have to go through the arduous process of figuring out which elements of a stretch of DNA are actually functional. To synthesize or "clone" a gene, the messenger RNA molecule is isolated and used as a template on which a new strand of DNA is formed. The strand that is synthesized in this way is called *complementary DNA* (cDNA). The cDNA can then be studied to determine which protein it codes for, or to locate this particular gene in other DNA. These strands are also called *expressed sequence tags*, or ESTs.

## Gene Chips

A creative new method of genome research combines the principles of base pairing with microchip technology to provide a means of rapidly identifying many strands of DNA at the same time. This technology is known as a *DNA chip*. Its creators have adapted microchip fabrication methods to place tens of thousands—or even millions—of single-stranded DNA segments onto the surface of a chip, thus providing a venue where multiple reactions can take place at the same time. The chip can be used to diagnose mutations in large genes that can be millions of bases long. It can also be used to monitor which genes are expressed, or active, in a particular type of cell, because it can monitor the activity of many genes at the same time.

To develop a gene chip, researchers divide a surface of glass or silicon into sites called *features*. At each feature they attach copies of a single DNA segment, which is a probe with a known sequence. They can range in length from several nucleotides to millions. The genetic material that will be tested on the chip is first cut into pieces using restriction enzymes. The resulting pieces are labeled with a fluorescent marker and the sample applied to the chip. The marked DNA sticks to the device only where it pairs up with a DNA probe whose sequence is complementary to all or much of its own.

---

### MAKING A HAYSTACK FULL OF NEEDLES: THE POLYMERASE CHAIN REACTION (PCR)

The principles of base pairing have been put to good use in the development of methods of duplicating a strand of DNA. In the early days of genetic research, one of the most frustrating barriers to studying DNA was the fact that there was frequently too little DNA to work with. This was particularly so when investigators began to use DNA testing as a way to identify criminals or the victims of crimes (see page 137). Their source of DNA might only be a few drops of blood or strands of hair. The *polymerase chain reaction* (PCR), a method of creating lots of strands of identical DNA, is widely considered one of the most important advances in DNA technology. Molecular pathologist Daniel H. Farkas describes PCR as a "biological reaction that generates a haystack full of needles." Its inventor, Kary Mullis, was awarded the Nobel Prize in 1993.

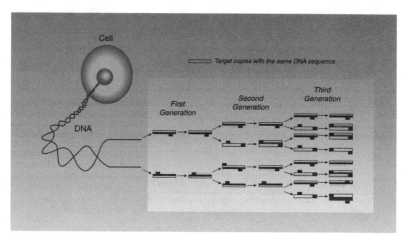

**Figure 5.2**   Using the polymerase chain reaction (PCR), a particular region of DNA present in a sample can be amplified so that millions of copies can be generated in just a few hours.

PCR works by using one strand of DNA as a template for rebuilding its complementary strand. First the DNA being studied is heated so that the double helix is denatured, or splits apart. Next it is cooled and short DNA oligonucleotides, or primers, are attached to the two ends of the DNA sequence that needs to be duplicated, called the *target sequence*. The mixture is then heated again along with a solution consisting of the four chemical bases that make up DNA and the enzyme DNA polymerase. This encourages the bases to bind, or hybridize, to the primers and to one another to regenerate the complementary strand. The mixture is then reheated and the whole process is repeated again and again. With each cycle, the amount of DNA in the entire sample is doubled, and millions of copies of the target sequence of DNA can be created. Because PCR requires alternate cycles of heating and cooling, one of the initial problems was the need to keep renewing the enzyme, which was destroyed by heat. The discovery of a bacterium, *Thermus aquaticus*, that flourished in the hot springs of Yellowstone National Park solved the problem. Taq Polymerase, the enzyme purified from this denizen of hot springs, has become the workhorse of PCR.

## Friend and Foe: Bacteria, Viruses, and Retroviruses

When studying and manipulating the stuff of life, it makes sense to start with the simplest living organisms. Thus, many of the most important past and continuing developments in genetics and genetic engineering involve such microscopic organisms as viruses, retroviruses, bacteria, and yeasts.

Bacteria were the creatures used in many of the first experiments in genetic recombination, and they continue to be the producers of a wide range of genetically engineered products. One of the best-known bacteria is *Escherichia coli*, or *E. coli*, a resident of the intestinal tract. It has been used so extensively by scientists that its microbiology and genetics are today better understood than that of any other organism. It is also one of the organisms most used in genetic engineering, as is Lambda, a bacterial virus—or bacteriophage—that makes its home inside the *E. coli* bacteria. (Lambda was one of the first organisms to have the complete sequence of the bases of its genome determined. It contains exactly 48,514 bases, which code for about fifty proteins, most of which are used to make the outside coating of the virus itself.)

In genetic research and engineering, microbiologists take advantage of a characteristic that makes these microorganisms scientifically fascinating, often useful, and medically challenging. These microorganisms reproduce rapidly and also mutate readily, changing the genetic sequence of their DNA. In some cases, bacteria mutate by incorporating genetic material from their hosts. Geneticists believe that early bacteria, which were among the first living organisms on earth, were able to survive in the hostile environment of our young planet and make evolutionary progress because they could repair damage to their genes by incorporating genes from neighboring microbes or use enzymes to make a fresh copy. The process of incorporating new genetic material into a bacteria is known as *transformation*. This kind of movement of genes from one cell to another does not happen in human, animal, or plant cells, where the genetic material of the cell is enclosed in a nucleus. This ability to "hijack" DNA is taken even further in viruses and retroviruses. The genetic material of the latter, one type of which causes AIDS, is actually composed of RNA. Once it infects a cell, an enzyme in the retrovirus called *reverse transcriptase* assembles viral DNA in the host cell using the RNA as a template. Another enzyme then splices it into the cellular DNA. Using the integrated DNA as a blueprint, the cell makes viral proteins.

The tendency to mutate is the reason why harmful microorganisms become resistant to antibiotics. As these microorganisms rapidly reproduce, some of them develop chance mutations that render them immune to a particular antibiotic that successfully killed off its ancestors. The bacteria may, for example, assemble or incorporate a gene for an enzyme that blocks the action of the antibiotic. To combat the possibility of successful mutation by the harmful bacteria, doctors strongly urge patients to take full courses of antibiotics, so that all the infecting bacteria are wiped out before they can develop mutations. Another way to counter the development of harmful mutations in bacteria and viruses is to use a "drug cocktail" strategy. Doctors reason that if they use a combination of drugs to attack disease-causing microorganisms, there will be less chance that they will develop a dangerous mutation. Researchers engaged in the ongoing battle to vanquish the rapidly mutating retrovirus that causes AIDS, for example, have attained the greatest success with combinations of drugs that attack it in different ways. This strategy increases the chances of successfully disabling the retrovirus as well as its various mutations.

To insert a foreign gene into target DNA, genetic engineers make use of the ability of microorganisms to take up new genetic material and to insert it into the larger DNA of a host. The microorganisms used are usually viruses or plasmids, which are circular pieces of DNA that exist outside the chromosomes of bacterial or other cells. First the new genetic material is inserted into the plasmid. To insert a foreign gene, the plasmid is cut by a specific restriction enzyme and the foreign gene is spliced in using another enzyme, called *DNA ligase*. The plasmid is next inserted into the bacterium, where it replicates. To ensure that the transformation has taken place, researchers usually insert some kind of "marker" gene as well. One trick is to insert a marker gene into the plasmid, which confers resistance to an antibiotic. Then only the bacteria that has received a copy of the plasmid that contains both the new gene and the gene for antibiotic resistance will be able to grow in a culture if the antibiotic is present. Another "marker" is to insert a gene coding for the ability to digest a certain nutrient. If the nutrient is the only one available, then non-engineered bacteria will die.

## From Genes to Kiwifruit and Dorset Sheep: The Many Facets of Cloning

In sexual reproduction, the genetic material of the male and female of a particular species combine so that the offspring have a mixed genetic heritage. A clone, on the other hand, is produced by asexual reproduction, and its genetic composition is identical to that of its parent. In the context of

## CELLS IN PLANTS, ANIMALS, AND MICROORGANISMS

Cells are defined as the basic unit of life, and the differing structure of cells of plants, animals, and microorganisms has important implications for cell cloning and genetic engineering. Plant and animal cells have a nucleus (which contains most of the genetic material), and are called *eukaryotes*. Organisms, like bacteria, that do not have a nucleus are called *prokaryotes*. All animal cells are differentiated; that is, they have a full complement of DNA, but the genes that do not contribute to the functions of the particular cell, like skin or muscle, are "turned off." Mammal cells, if cloned, can only generate cells of the same tissue type unless they are subjected to special conditions. (Some mammalian cells, such as nerve cells, may not divide at all after a certain age.) Plant cells are less differentiated, and most multicellular plants contain groups of cells that are *totipotent*: they can generate an entire plant when treated with certain plant hormones. Thus, plant cloning was successfully carried out well before the first successful cloning of a mammal from an adult cell was reported in 1997. In fact, the world's entire crop of kiwifruit plants has reportedly been cloned from the same parent plant.

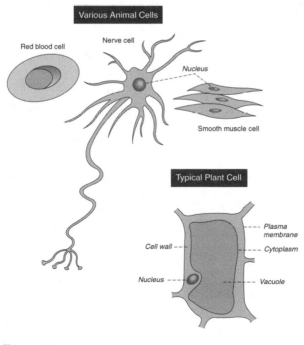

Figure 5.3a–e

biotechnology, cloning refers to a number of different procedures, not just the asexual production of an organism. Genetic researchers and engineers often talk about cloning genes. This refers to the mundane process of making many copies of transgenic bacterial or yeast genes that have incorporated foreign DNA into their genomes. As well as being used to produce numerous copies of a transgenic genome—of a yeast, bacteria, or plant—cloning is also used to decipher base pair sequences of genes that have not been modified. Once the nucleotide sequence of a gene has been identified, the gene is cloned or copied using base pairing principles, so that the gene, and the protein that it codes for, can be studied. The resulting database of information that is developed for multiple genes is known as a *gene library*.

In plant cloning, a genetically identical copy of a plant is produced asexually by various methods. Some plant cells, especially at the growing tip, are totipotent, which means they can produce all the various parts of a plant. To clone the plant, one can cut the shoot, apply growth hormones to the shoot, and develop a new plant. The same process can be applied to plant cells that have been genetically engineered to produce plants that have new genetic traits. At this point the cloning of plants or plant cells is fairly unremarkable because of the existence of totipotent cells in mature plants.

Cloning animals from mature cells is far more difficult. In fact, methods of successfully cloning animal cells to generate a living creature that will grow to be an adult are still nascent. That is why the reported cloning of Dolly the sheep in 1997 created such a stir in scientific circles and among the general public. The researchers who cloned Dolly claimed that she developed from the udder cells of a six-year-old ewe. This was viewed as an astounding scientific accomplishment because until then it had been impossible to clone an animal from cells that were even a few weeks old. In the 1980s, researchers successfully cloned sheep and cows from embryonic cells that had not yet become specialized, but mature animal cells are differentiated. This means that if a skin cell divides, it produces another skin cell, but it cannot be prodded to produce bone or blood vessels. Although every cell in an animal's body has the full complement of DNA that codes for every protein involved in the animal's life processes, the genes that are not needed in a particular type of cell's function are switched off, or disabled.

To clone an animal successfully from a mature cell, scientists have to find a way of turning differentiated genes back on. The Scottish researchers who cloned Dolly achieved their aim reportedly by starving the cell, and then inserting the nucleus into an ovum, or egg cell, which had

had its nucleus removed. Researchers believe that the egg cytoplasm that surrounds the nucleus contains as-yet-unidentified proteins that can reprogram genes to switch on and reproduce the various tissues of an entire animal.

Scientists are anxious to develop reliable methods of cloning animals from adult cells because it is a faster and more predictable method of reproducing an animal that has desirable genetically engineered traits. Transgenic animals, whose genetic heritage includes foreign genes, are very valuable because they can be used to study human diseases and engineered to produce valuable proteins. Such animals are used to model the course of a human disease, like cancer, and to test the efficacy of drugs before they are taken into human trials. The first transgenic animal to be patented, in 1988, was Oncomouse, a cancer-prone mouse that is widely used to test new cancer drugs. In the future, transgenic animals may be used extensively to produce medically useful drugs and proteins, which cannot be produced by bacteria. The blood of a transgenic goat, which contains human protein, could be suitable for human blood transfusion. The milk of transgenic goats is already being used to produce blood-clotting factors to treat hemophiliacs. There is also research into using transgenic pigs, engineered to have human surface proteins in their hearts, for organ transplants.

If an animal is genetically engineered to produce a valuable drug in its milk, cloning is a quick method of reproducing more of those animals and ensuring that the particular trait is passed on. It is also easier to insert useful genes into an animal cell and then clone those cells so that all of the cells in the resulting animal have the desired gene. If cloning is not used, the only way to genetically engineer sheep and cows is to inject a gene directly into a fertilized egg. This method is not reliable. Usually a researcher has to inject thousands of fertilized eggs to get one transgenic animal that lives into adulthood.

For more than one year after a Scottish researcher reported the successful cloning of Dolly from the mature udder cells of a ewe, scientists were unable to repeat the feat, a development that has raised questions about whether Dolly was, in fact, a bona fide clone. Dolly's mother was pregnant when cells were taken from her udder. In some animals, fetal cells enter the mother's blood and could possibly get into the udder, so an undifferentiated fetal cell could have been Dolly's parent. In fact, in May 1998, American scientists reported the successful cloning of three Holstein calves, using a different method. Those investigators inserted genetically altered fetal cells into the nuclei of unfertilized cow eggs whose own genetic material had been removed. Also in 1998, an international team of researchers at the University of Hawaii reported the successful cloning of dozens of mice.

Although, at this time, reports of successful animal cloning are few and far between, most observers believe that in the future, scientists will routinely accomplish the successful cloning of animals. They also believe that the same technology for cloning animals will eventually be applied to human beings. This prospect is highly controversial. The public is subjected to a daily diet of sometimes sensationalist news on advances in genetic engineering. Reports of Dolly's cloning immediately raised the specter of a thousand Elvis Presley clones and of hatcheries producing thousands of identical humans, each destined to fulfill a predetermined role in life. The prospect of human cloning has also been viewed positively by some. Parents who have lost a child in an accident, or whose child was severely brain-damaged in an accident or a hospital mishap, may welcome the chance to produce another child with the same genetic makeup. There is also the prospect of literally resurrecting important historic figures like Albert Einstein or Abraham Lincoln, who, because of their personality and intellect, have made great contributions to mankind.

In reality, batches of human clones are unlikely to be produced anytime soon. The technology is still developing. More than fifteen months after her appearance, Dolly had no fellow clones. Moreover, she was created after 277 failed attempts, a ratio that is unlikely to be acceptable—or affordable—in human production. But as reproductive technology advances, the birth of human babies who are genetically identical to one parent cell may become a possibility. At this stage, women who are in their late forties and even their fifties can successfully bear children from donated ova that are fertilized in a petri dish and then implanted in their wombs. In fact, the chances of an older woman having a successful pregnancy in this way, rather than using her own "old" eggs, are actually significantly higher. Similar technologies may be used in the future so that a woman can gestate an implanted embryo that has been successfully cloned from one parent cell.

In the interim between possibility and realization, many serious ethical issues need to be addressed. There are concerns that people will be "bred" to have desirable attributes, and then cloned to ensure that these attributes are perpetuated. People who do not have the benefit of this breeding may be discriminated against. There is also the question of whether cloning Albert Einstein's cells will actually result in the same person. Identical twins are technically clones. Although they can often be very similar, differences in personality, accomplishments, and interests can be observed in identical twins even in childhood. We do not know how a person's genetic endowment interacts with his or her parenting and education. Transplanted into the twenty-first century, Albert Einstein's genes may not actually produce Albert Einstein.

There are also technical considerations that have not yet been addressed. One caveat is the possibility of genetic defects in the cell that is being used for cloning. As a cell ages, DNA mutations accumulate. Mutations that do not affect the cell's current function may not be an issue unless the cell's dormant genes are switched on when it is cloned. Generally, an embryo that has serious genetic damage will die; however, there is the possibility that moderate genetic damage will not prevent the embryo from maturing. Then, in future years, the human being cloned from that cell may suffer from serious disabilities. Another caveat is the possibility that a human clone will have a higher chance of having defective DNA because it will have literally been "purebred." It is almost an article of faith that mixing the gene pool, which is what happens in sexual reproduction, is beneficial. Every human genome carries deleterious genes that cause diseases, raising the question of how scientists can ensure that "bad" genes do not get passed on in the process of cloning. To clone a human successfully, one has to assume that genetic engineers can "fix" deleterious mutations in the human genome, an enormous and potentially unachievable undertaking. To accomplish this, it will be necessary for geneticists to have an encyclopedic knowledge of the human genome, a huge research project that is currently under way.

## Spelling Out Our Molecular Legacy: The Human Genome Project

One of the ongoing goals of genetics researchers is to sequence the DNA of many living organisms and to use that knowledge to understand the repertoire of proteins that genes make, and how these proteins orchestrate the interdependent life processes of a living cell. By 1998 the genomes of various microscopic organisms, including several viruses, bacteria, and yeasts, had been fully sequenced. Some geneticists believe that ultimately they will be able to use this knowledge to develop a "periodic table" of gene families that will reveal a natural order to the science of genetics and enable further discovery and research in the same way that the periodic table of the elements informs chemistry (see chapter 8).

At the other end of the genetics spectrum is a massive and ongoing study of human DNA known as the Human Genome Project, which is probably one of the largest and most far-flung scientific projects ever attempted. Packed into a nucleus less than a hundredth of a millimeter across, the entire bundle of human DNA is known as the human genome.

## CHROMOSOMES

All large genomes, like ours or those of plants and other animals, are organized into distinct, separate, microscopic units called chromosomes. Genes are arranged along the chromosomes in a line. The nucleus of most human cells contains two sets that consist of twenty-two non-sex autosomes and an X or Y sex chromosome. (A normal female will have a pair of X chromosomes; a male will have an X-and-Y pair.) Chromosomes contain roughly equal parts of protein and DNA.

Chromosomes can be seen under an optical microscope, and when stained with certain dyes, they reveal a pattern of light and dark bands reflecting variations in the amounts of bases (A and T versus G and C) they contain. The chromosomes can be distinguished from each other because of differences in size and banding patterns. A few types of major chromosomal abnormalities, including missing or extra copies of a chromosome or gross breaks and rejoinings (translocations), can be detected by microscopic examination; Down's syndrome, in which an individual's cells contain a third copy of chromosome 21, is diagnosed by chromosomal analysis. Most changes in DNA, however, are too subtle to be detected by this technique and require molecular analysis.

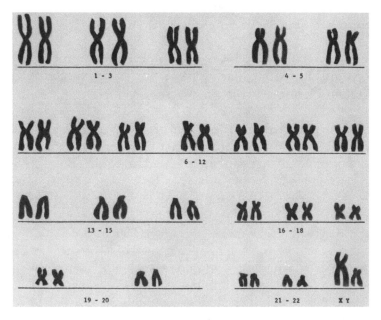

**Figure 5.4**  A set of male chromosomes, numbered to reflect their distinctive genetic makeup and location in the nucleus. Note the XY sex chromosomes. Women have two X chromosomes instead.

In human sperm and egg cells, it consists of 3 billion base pairs strung in varying combinations to form an estimated 80,000 active human genes as well as vast stretches of genetic material that do not as yet appear to have any function and are termed "junk" DNA. (Somatic, or non-sex, cells are equipped with two sets of genes, and thus contain 6 billion base pairs of DNA.) It is widely predicted that the findings of the Human Genome Project will have immediate or potential use in diagnostics and possible future therapies. Biologists expect that the gene maps and sequence information generated by genome research will be used as a primary information source for human biology and medicine far into the future.

Internationally, thousands of scientists and technicians are engaged in sequencing the codes of our DNA. In the United States, the project started under the umbrella of the National Institutes of Health (NIH), which had planned to complete the sequence by the year 2005, after a fifteen-year program costing $3 billion. Under the auspices of the NIH, six university laboratories are working on the sequence, each studying a different chromosome. Other laboratories around the world are also deciphering the human code. Almost every day these laboratories place thousands of newly deciphered genetic sequences onto the Internet, from which the information is incorporated into the databases of other researchers. Researchers expect the project to be completed by the year 2005 or possibly even earlier, as research methods improve and the cost of gene sequencing decreases. By the middle of 1998, the Human Genome Project had completed sequencing about 3 percent of the human genome. At that time, Perkin-Elmer, an American company that manufactures scientific instruments, and Craig Venter, a molecular biologist who heads a genetics company, announced their intentions of sequencing the human genome, using at least two hundred automated gene-sequencing machines to do so. The sequencing methods used by members of the Human Genome Project are quite labor-intensive. Automating the task is expected to speed up the process because the new generation of sequencing machines can work around the clock with very little attendance. Venter and Perkin-Elmer estimated that they will be able to complete sequencing the human genome by 2001 at an estimated cost of only $200 million.

In addition to sequencing the human genome, Venter and Perkin-Elmer plan to construct a database that will integrate medical and other information with the DNA sequences. One important element of the new database will be to elucidate common variations in human DNA, known as polymorphisms. These are the differences that make each individual unique, even though all people and ethnic groups are believed to have mostly similar sequences of DNA bases in their genome.

Venter, who is known for developing innovative research strategies, is using an unorthodox approach to decode the human genome. He plans to use a method that is known as "whole-genome shotgun sequencing." This strategy involves blasting an organism's DNA into thousands of fragments. Automated sequencing machines then read the base sequences at the ends of each fragment. This process is done with many copies of a genome. Overlapping sequences are then matched by computer, and the full genome is then assembled. The international genome project takes a much more laborious approach. Its strategy is known as the clone-by-clone method. First the genome is divided into small fragments that are copied, or cloned, by being incorporated into bacteria that then divide, making many copies. These clones are then sequenced, but investigators first try to figure out, in a process called mapping, where on the chromosomes they belong. The sequencing and mapping is painstakingly checked and rechecked, a finishing phase that is the most expensive and time-consuming.

*Science News*, a respected weekly news magazine, describes the contrasting approaches of the Human Genome Project and Venter's group as a race between the tortoise and the hare. At this time the outcome of the race cannot be predicted, but one probable result is that the more rapid sequencing of the entire human genome will hasten the growth of a new field called post-genomics research. The aim of this challenging research will be to interpret the voluminous information that will be generated as the lengthy gene sequences of humans and other animals are revealed. Because of the size of the human (or any complex animal) genome, geneticists have likened mammal genome mapping to producing the map of a country that gives the location of each house. One estimate of the vast amount of data produced is that, if compiled into books, the sequenced human genome would fill an estimated two hundred volumes the size of a Manhattan telephone book. Indeed, skeptics question the utility of this project, contending that the vast amount of information generated by genome mapping projects will ultimately serve little purpose, particularly because in the case of the human genome, a large portion of the DNA appears to have no function at all.

An influential critic of the government approach is William Haseltine, head of Human Genome Sciences, which does gene research. "Imagine that the genome is an encyclopedia with about three billion letters," he wrote in an opinion piece in the *New York Times*. "Buried within this text are about 100,000 sentences (the genes) that tell the body what essential proteins to make. The sentences are separated from one another by page after page of random letters—what scientists call junk DNA. To make matters even more complicated, the sentences themselves are also fragmented and interrupted by pages and pages of random letters—more junk DNA. In fact, less than 5 percent of our DNA contains real information. The other 95 percent has no genetic meaning."

Haseltine claims that his group has already spelled out the sequences of almost all of the functioning genes in human DNA. His group used sequencing methods that "read off" gene sequences from messenger RNA (mRNA) that has already edited out the noncoding DNA from a particular gene.

Haseltine's critique might have merit if efficiency were the only concern of genome researchers. However, there are those who disagree with his contention that all noncoding DNA is "junk." Some researchers believe it may play a role in orchestrating the activity, or expression, of genes. Other researchers have found that repeated patterns of noncoding DNA are very useful aids when they have been searching for a particular gene. Some patterns of noncoding DNA are inherited along with a gene. When researchers study the DNA of family members who are all affected by a certain genetic disease, they have also found patterns of repeated junk DNA that are frequently located near the gene they are looking for. Junk DNA is also used in forensic research (see page 137).

## ELECTROPHORESIS AND BEYOND: HOW THE SEQUENCES OF GENES ARE DECIPHERED

To sequence a stretch of DNA rapidly, researchers use automated sequencing machines that are based for the most part on methods developed in the 1970s by Harvard University researchers Walter Gilbert and Allan Maxam, and by Fred Sanger in Cambridge, England. The Maxam-Gilbert method uses chemicals to break the DNA into fragments at a specific base, such as adenine. Other chemicals cut the fragment at another base region, such as guanine. The resulting fragments are placed onto a gel in different channels depending upon the chemical used, and an electric current is applied. The DNA fragments are attracted by the electrical current and move through the gel at different speeds depending upon their size. Because each "cut" results in a DNA fragment of a different length, the resulting sequence can be read off from the position of a particular fragment.

Rather than breaking up DNA, the Sanger technique uses enzymes that make new DNA chains complementary to the strands of DNA that are being sequenced, but that stop at a certain point of the strand. First the DNA strands are separated by heating. One strand is saved, and thousands of copies are made using PCR. The DNA sample is then divided into four portions and placed into a test tube. Each test tube contains DNA primers that hybridize by base-pairing to their complementary sequence in the DNA. The solution also includes all four of the DNA bases, as well as DNA polymerase, which is an enzyme that assembles the DNA bases and attaches them to a complementary strand of DNA. Each test tube also contains a different modified DNA base. When each particular base attaches itself to the newly synthesized DNA, it stops the synthesis. Thus, modified base A stops the synthesis at an adenine, modified base T, in another test tube, stops the synthesis at thymine, and so on. At the end of the process, thousands of reconstructed strands will be in each test tube. All of the strands have the same sequence of bases, but they terminate at different positions depending upon where they linked up with a modified base. The four samples are then placed onto different channels in a gel, where the principles of electrophoresis take over. The DNA fragments travel to different positions depending upon the length of the fragment. The sequence of bases can then be deduced by reading their position on the gel.

Another variation of the Sanger technique uses fluorescent dyes to tag the modified bases. Then, instead of dividing the sample into four, the same test is done in one test tube. The resulting fragments are still run through a gel, where a laser reads off the bases by color. The resulting chart looks something like the kind of graph an electrocardiogram would generate, with each color coming out in peaks and troughs. The highest points of the graph mark the termination of a strand, and that is the point from which the base sequence is read.

These methods of sequencing a stretch of DNA are inferential; the DNA sequence is deduced by indirect means. In the future it may be possible to directly read the base sequence of a DNA strand using a scanning tunneling or atomic force microscope.

Among the first targets of research into the human genome has been the study of "single gene diseases," in which a single defective gene is the cause of disease. Once it has been discovered, researchers give the gene a short name, called a "tag," that often indicates its function; thus BR(east)CA(ncer) I for the first discovered gene that is strongly linked to hereditary breast cancer. A gene becomes defective if the sequence of its bases is changed so that one or more amino acids of the protein coded for by the gene are altered, and as a result, the protein can no longer do its job. The DNA damage can be drastic, like a deletion in which all or part of a gene is eliminated, or as subtle as the change in a single base. For example, phenylketonuria, an inherited disease that severely reduces mental ability if untreated, is due to the change of just one base in the gene coding for a protein containing 451 amino acids. This kind of research provides valuable information about the role particular proteins play in human metabolism. Research into genetic diseases also helps doctors and patients make informed choices. For example, once diagnosed, a "single gene" inherited disease like phenylketonuria can be managed by careful attention to diet. Prenatal diagnoses of fatal genetic diseases, like the neurological disorder Tay-Sachs disease, enables parents to make decisions about whether to end a pregnancy, or to make appropriate medical plans should they decide to carry the pregnancy to term.

New diagnostic tools provided by DNA research are already in use to identify diseases, plot the progress of conditions like cancer, pinpoint pathogens, and find compatible donors for organ or bone-marrow transplants. DNA chips, with their capacity for large numbers of probes, are finding increasing applications in the analysis of complex genes and genetic diseases. These developments are leading to the growth of a new medical specialty: molecular pathology. Rather than identifying a particular disease-causing organism under a microscope, the molecular pathologist will instead make a more accurate diagnosis based on DNA sequences. Faced with the challenge of finding donors for bone marrow transplants, for example, one Canadian company has developed a technique to sequence a three-hundred-base-pair stretch of DNA that research has established is involved in orchestrating the body's immune response to invaders. Using restriction enzymes, researchers pick out this particular stretch of DNA from a sampling of the patient's cells. An automated sequencing machine then analyzes the base pair sequence of the sample. A computer program then checks established gene banks for a possible match. This kind of "tissue typing," which formerly took days to accomplish, is now done in a matter of hours.

One area in which it is hoped that genetics research will lead to breakthroughs that until now have been elusive is the prevention and treatment of cancer (see chapter 6). At this stage, cancers are widely considered to be diseases of the genes—either caused by a genetic predisposition or by a cascade of mutations during a person's life span. One of the major assumptions in cancer treatment is that early diagnosis, made before the cancer cells can spread, or metastasize, can save lives. At present, most cancers are diagnosed when the malignant cells have formed a tumor, which can already consist of millions of malignant cells. With genetic-based diagnostics, it may be possible to diagnose and forestall the spread of a cancer well before the tumor stage. For people who carry an inherited predisposition to a certain type of cancer, a genetic diagnosis could facilitate more effective monitoring. In addition, in the future, when researchers have teased out the complex relationship between a genetic predisposition to a cancer and such environmental factors as a particular toxic chemical, a person with a diagnosed susceptibility to a cancer may be able to avoid those environmental triggers. Important findings include the identification of two unrelated mutated genes, BRCA I and BRCA II, the first that clearly appear to strongly predispose women carrying the mutation to breast and ovarian cancer.

However, the nature of BRCA I and BRCA II also point to the immense complexity of teasing out the relationship between genes and cancer. Not only are a very small percentage of breast cancers inherited, but mutations in BRCA I and II appear to play a role in a widely varying percentage of inherited malignancies, depending on family history. Moreover, both are large genes, and hundreds of mutations have turned up all through them. These mutation variations make it hard to track how the gene is inherited, which mutations are actually inherited, and which mutations render the protein this gene codes for useless. Moreover, too little is known about the mutations to give the women carrying them a precise estimate of what they mean.

Genome research has already stepped onto the thorny grounds of commerce, with one Utah company offering a high-priced test that sequences BRCA genes to scan for 400 mutations. A number of universities offer a more limited set of tests. But many researchers caution that these kinds of tests are extremely premature and potentially even dangerous: being screened for a specific mutation does not mean a person won't develop cancer, and not every mutation associated with breast or ovarian cancer can be detected. Moreover, one-third of all breast cancers that seem to run in families don't show any linkage to either BRCA I or BRCA II. As in many genetic diseases, epidemiological studies are an indispensable research tool. In the United States, the National Cancer Institute has organized a Cooperative Breast Cancer Registry, which will collect blood cells and tumor specimens from about six thousand women with breast cancer and twenty thousand of their relatives. These samples will be studied for mutations in either BRCA I or BRCA II, or other as-yet-undiscovered "breast cancer" genes. In these kinds of studies, researchers use a method called "positional cloning," which seeks out "the needle in the haystack" by looking for common patterns in the DNA of families or closely related populations of people known to be affected by a specific disease. When the common patterns, which might be repeated sequences of "junk" DNA, are identified, researchers then attempt to locate the specific gene responsible. They assume that the gene will be located nearby because of the familial pattern of inheritance shown in the "junk" DNA.

## Nature and Nurture:
## The Interplay of Genes and the Environment

Although there is a clear link between a genetic mutation and disease in "single gene" diseases like phenylketonuria, the relationship between genes and many other diseases is less clear. Genetic variations that appear to predispose a person to a particular disease are called *susceptibility genes*, and generally they are affected by a host of other genes as well as factors in the outside world. One approach to preventing or treating diseases in carriers of susceptibility genes is to track down the environmental factors shaping the effects of the gene. In 1994, for example, British researchers found the first susceptibility gene for asthma and related allergic diseases. This gene turned out to encode part of the receptor that responds to immunoglobulin E, an antibody that triggers allergic responses. Now several groups of researchers are closing in on other asthma-susceptibility genes. It is hoped that researchers will be able to link particular gene variations in asthmatic people to the environmental exposures that set off their attacks. Based on the assumption that it is easier to change the environment than to change a person's genes, the best strategy will be for people to avoid these triggers in childhood, when it is believed that lifelong allergic reactions are probably first established.

Human genome research has also raised weighty social and moral issues, with fears that insurance companies and potential employers will discriminate against people diagnosed as carrying deleterious genes. There are also concerns that parents will take their quest for perfect children too far, selecting genes for appearance and intelligence and rejecting fertilized embryos (at even the single-cell stage) which indicate the potential for disease or imperfection. There are several notable historical figures whose deleterious genes would certainly have disqualified them if perfection was the goal: It is widely believed that Abraham Lincoln, for example, was a victim of Marfan's syndrome, a genetic condition that, besides causing people to be tall, with long legs, arms, and fingers, can also cause life-threatening heart problems.

Worldwide, there is a rush to patent human gene information as it is deciphered. In the United States, the U.S. Patent and Trademark Office has already issued thousands of patents on human gene sequences. Critics feel that this kind of activity stifles biological research by placing information and research methods off limits to those who cannot afford to pay licensing fees. Pharmaceutical and bio-informatics companies contend, on the other hand, that without financial incentives, the costly and laborious labor of genome research would not get done.

There are also other factors to consider: Even when every gene in the human genome has been sequenced, diseases that are caused by genetic mutations will still persist because the DNA in human cells continues to mutate in a living person. Another consideration is that because everyone carries deleterious genes, perfection is impossible. Many diseases involve the interplay of several genes with environmental factors. The processes of these genetic events have only begun to be understood. At the molecular level, the physiological processes of any living cell are astoundingly complex, involving cascades of gene expression in a mysterious interplay with the cellular environment. It may be that the intricate choreography of DNA will never be totally understood and can never be completely controlled.

## Pharmaceuticals of the Future: Genome Research Fuels Drug Discovery

Advances in genome research can enable the development of a new generation of pharmaceuticals; cognizant of this, drug companies are some of the biggest players in DNA research. More precise genetic information about viruses and bacteria can be used to develop drugs that operate with delicate precision rather than with the scattershot method of the broad-spectrum antibiotic that kills off a range of microorganisms, including those that can be beneficial.

DNA codes can be deciphered and manipulated to develop a large number of antibodies that will help boost immune system responses, or help deliver a potent drug more directly to its target. There is reason to hope that DNA research will facilitate the difficult and costly development of drugs that fight viruses and cancer (as well as elucidating more precisely the environmental factors that are responsible for triggering the DNA mutations that cause cancer). Whereas conventional chemotherapy acts by killing off all cells in the body that are rapidly dividing, a new DNA-based generation of cancer therapeutics may act by disabling key elements of a particular gene, or may be specially tagged to recognize and kill malignant cells. Genome research is also an important element of two new methods of drug development that use combinatorial chemistry and rational drug design (see chapters 6 and 8).

Growing knowledge of the human genome is also enabling the development of a new field called pharmacogenomics. This will be the tailoring of pharmaceuticals to the particular genetic makeup of a patient. This field is important, because individual patients can have subtle genetic variations, called polymorphisms, that make them either unresponsive or particularly susceptible to dangerous side-effects from certain medications.

## Doctoring DNA: The Promise of Gene Therapy

If the cause of many diseases lies in the genes (either those of the sufferer or of the organism that has infected the sufferer), then the most effective and efficient way to cure the disease is to "fix" the gene by finding a way to insert the "normal" gene into the sufferer's DNA. (A related approach is to insert DNA into the DNA of damaging organisms or cells that kill the intruders or make them more vulnerable to certain drugs.) This treatment is called gene therapy, and it also holds the promise of curing diseases of genetic origin, such as Tay-Sachs disease or cystic fibrosis and some cancers, that so far cannot be cured any other way. Theoretically, then, instead of treating an adult diabetic with insulin injections, gene therapy will insert the gene that manufactures insulin into the cell.

That's the theory. The practice is far more complicated, with hurdles that some believe may never be overcome. Although genetic engineers have a wide variety of tools, these methods do not always work with precision in complicated animal cells. The genome, noted one researcher, is not like a library with orderly shelves of books, but more like a "chaotic archive with the genetic material twisted together into an extremely cramped space." Hurdles include the challenge of delivering the gene through the cell wall and into the cell nucleus while ensuring that the body's immune system does not recognize the foreign DNA as an intruder and destroy it. Once inside the nucleus, researchers must devise a way to insert the gene in the right position in the DNA, and ensure that the genes that promote expression of the desired DNA are included in the transfer.

Then there is the challenge of delivering the gene to enough cells to ensure that the required enzyme or protein is manufactured in sufficient quantities, and that the gene is delivered to the organ where it is required. Some genes also need careful regulation so that they manufacture the right amount of the necessary protein at just the right time. That kind of control is difficult to achieve. Inserting genes into human embryo or germ line cells, which would then divide with the gene in place, and be passed on to future generations, is an option that so far has been ruled out for ethical reasons. The first single-gene disease that has been treated with gene therapy with moderate success is ADA deficiency, a disorder in which the mutation of the gene coding for the enzyme adenosine deminase results in the destruction of the immune system. Gene therapy has also been attempted on patients suffering from cystic fibrosis, who showed some benefit for a limited period of time.

## Sneaking In: Methods of Delivering DNA

To deliver therapeutic genes to human patients, researchers are experimenting with various approaches. In *ex vivo* methods, researchers remove cells from a patient, add the desired gene in the laboratory, and then return the altered cells to the patient. Using *in vivo* methods, researchers attempt to introduce the therapeutic genes directly into the target tissue in the patient's body. One direct method of gene delivery is to use a virus "vector." This procedure utilizes the life cycle of a virus—which inserts its own DNA into that of the cell to replicate its genetic material—to deliver a desired gene into the target DNA. In the complicated world of genetic engineering, however, the "virus vector" needs to be carefully selected and "engineered." For example, a relatively harmless small virus will readily spread in tissue or cross the blood-brain barrier, but its size limits the number of foreign genes that can be inserted into it. In utilizing larger viruses, researchers need to take care that they remove the damaging genes of those viruses that cause disease. At the same time, they must ensure that they do not cripple them so much that they lose their effectiveness as an infective agent. One method of skirting such problems is to create *virosomes*, which consist of the therapeutic gene and elements of the virus that enable it to target and invade host DNA, incorporated into a liposome. These spheres, which have a hollow center, are considered a good candidate for delivering gene therapy and other drugs in general, because they can gain access to a cell without being degraded.

The choice of vector also depends upon the type of cell affected by the genetic disease. For example, cystic fibrosis affects lung tissue; Tay-Sachs disease results in damage to the nervous system; and Duchenne's muscular dystrophy affects the muscles. To deliver effective gene therapies, researchers need to select viruses that traditionally attack these kinds of cells and establish latent or persistent infections, where the host cell is not destroyed. Retroviruses are a good candidate, because they do integrate into the host genome. However, most retroviruses can only infect dividing cells. Ironically, one of the most promising subfamilies of retroviruses, which can infect neuronal cells that do not divide any more, are called the lentiviruses, and include HIV, which causes AIDS. By 1998

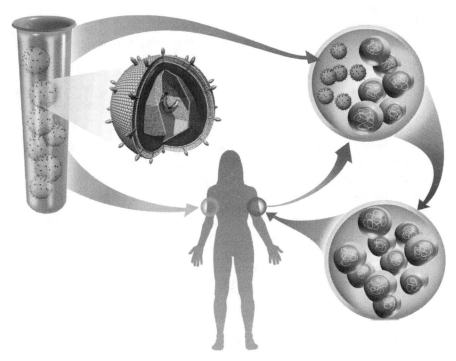

**Figure 5.5** Therapeutic genes are sometimes delivered to human subjects by injecting genetic material or agents (vectors) containing the material directly into the tissue (left arm). In another approach, physicians remove cells from a patient, add the desired gene in the laboratory, and return the corrected cells to the patient (right arm).

several retroviruses had been approved for gene therapy, to treat conditions ranging from brain tumors to severe combined immunodeficiency. To deliver long-term gene therapy, researchers are studying the use of viruses, such as herpes simplex, which persist as a separate entity (that is why some people continually get cold sores).

To avoid the problems inherent in harnessing viruses, researchers are also experimenting with non-viral delivery systems, which include lipoplexes. These are stretches of therapeutic DNA enclosed in liposomes. There has also been some success using strands of "naked" DNA that are not attached to a delivery system.

## The Gene of the Crime: DNA and Identity

One of the remarkable characteristics of DNA is its ability, under certain conditions, to survive for millennia. Using PCR (see page 117), scientists have been able to extract and amplify DNA from the frozen remains of a woolly mammoth that met its death about ten thousand years ago, and from an insect entombed in amber millions of years ago. A forensic geneticist can work similar wonders with the saliva residue on an envelope. Thus was Nidal Ayyad, one of the accused conspirators in the bombing of the World Trade Center in New York City, identified based on the PCR analysis of the saliva on an envelope mailed to *The New York Times* detailing his group's activities.

Most people are familiar with what traditional fingerprints look like. The DNA equivalent of a fingerprint, however, looks something like a bar code in which the bars are fragments of DNA of different lengths. DNA that is being analyzed for identification purposes is extracted from a sample, and then cut into pieces by restriction enzymes. The fragments are subjected to *restriction fragment length polymorphism* (or RFLP) testing. This testing works on the principle that while people have most of their DNA in common, there are areas particular to each person and each racial group that are distinctively different. This difference is reflected in the distinctive length of a person's DNA fragments when they are treated with restriction enzymes.

When a forensic geneticist studies a DNA fingerprint, the geneticist does not analyze the entire genome: the elements of DNA that form the "fingerprint" are not usually "functional" DNA, but areas in the "junk DNA" of a gene known as the introns. Genetic research has revealed that within specific areas of his or her DNA, each person has a variable number of repeat sequences, called *variable number tandem repeats* or *minisatellites*, and those differences are detectable. For example, the sequence AGCT may be repeated sixty-two times in tandem in one person and forty-seven times in another person. Forensic geneticists use many of the techniques of the genetic engineer, such as restriction enzymes, PCR, and synthetic DNA probes that are already known to hybridize (or link up) to areas of human DNA where these sequences are known to exist.

The typical process in forensic identification is as follows: Investigators take samples from a crime scene of blood, semen, or even a hair. The DNA is extracted, and, if the sample is small, amplified using PCR. The DNA is then chopped up with restriction enzymes and the pieces of DNA are separated by gel electrophoresis. In this process, DNA samples are placed into tiny holes or channels in a gel, and an electric current is applied to the gel. The varying-sized pieces of DNA, which are attracted to the electrical current, move at different rates. The size of any piece of DNA can be measured using markers, which are standard pieces of DNA of known size. The fragments, now sorted by size, are separated into single strands and transferred from the gel onto a membrane that fixes the fragments in place. At this point, DNA probes that are tagged with a fluorescent or light-generating enzyme are applied. The result is a complex pattern of bands that is unique for every individual. This pattern is then compared with a sample of DNA from a suspect or, in the case of paternity testing, from the alleged father. Companies that perform DNA fingerprinting have also developed databases based on the analysis of thousands of blood samples, and use this data to give probabilities of a particular DNA pattern appearing.

This time-consuming method of DNA fingerprinting has begun to be supplanted by automated systems that are capable of generating the base-pair sequence of a stretch of DNA (see page 129). In addition to analyzing DNA from the cell nucleus, forensic genetic analysis is sometimes done using DNA extracted from the mitochondria, of which there are several hundred in the cytoplasm, or body of a cell. Mitochondria are the cell's energy-producing power plants. The DNA inside them is inherited only from the mother, and serves as a perfect identity marker for maternal relatives. In old crime samples, like decayed bone, the DNA sample will most likely be mitochondrial DNA, since the cell nucleus may have already degraded.

PCR, the process of amplifying exceedingly small samples of DNA, has been a boon to forensic geneticists, enabling them to develop samples from minute amounts of evidence—a hair, a drop of saliva, or a tiny drop of blood. Forensic geneticists have used amplified DNA to identify the victims of airplane crashes, even in one case where bodies were submerged underwater for weeks. The amplified DNA is matched against a sample of DNA taken, for example, from stray hairs in a hairbrush. But PCR can also prove to be a bane because a sample subjected to PCR amplification can be easily contaminated. Thus, challenges to DNA fingerprinting in court are often based on questioning the methodology of the laboratory doing the testing. This strategy was used to devastating effect in the murder trial of O. J. Simpson, where the accuracy of the laboratory work performed by the Los Angeles Police Department was brought into serious question by Simpson's lawyers.

However, Harlan Levy, a former homicide prosecutor in New York City, considers DNA analysis to be the "greatest friend of the falsely accused." And indeed, in recent years, several convicts who have spent years in jail convicted for rapes they did not commit have been freed on the basis of the genetic analysis of evidence that is many years old.

## Down on the Farm: Genetic Engineering and Agriculture

Farmers have practiced a form of genetic manipulation since the dawn of agriculture, selecting plants and animals that appear to have sought-after characteristics and attempting to incorporate these traits in future generations. Thus the farmer of a traditional crop will select his best plants to produce seed for future crops. Dairy farmers, using judicious cross-breeding, developed specific breeds of cows that were superb milk producers. The price a racehorse fetches is based directly on his bloodlines: the more winners in his family tree, the higher the price.

Traditional breeding methods were laborious and expensive: The breeder would have to wait for generations for a sought-after characteristic to manifest itself, and then cross-breed repeatedly to get rid of undesirable characteristics. Modern plant breeding methods are also time-consuming and expensive. A plant breeder looking to incorporate a disease-resistant trait into corn, for example, must wait years for cross-pollination to produce the desired trait.

Modern genetic engineering in agriculture differs from its traditional predecessors in several key ways. The engineer goes directly to the DNA, and does not necessarily have to wait while an animal or plant goes through generations of reproductive cycles. The traditional breeder made selections based on visible traits, like the type of hair or horns an animal has, while the genetic engineer operates at a more subtle level, looking to produce more of a specific protein, for example. Most significantly, the genetic engineer combines the DNA of different species. The Union of Concerned Scientists compiled the following table illustrative of some re-combination field trials that are currently under way:

| Crop | Source of New Genes | Purpose |
|------|---------------------|---------|
| Potato | Chickens | Increased disease resistance |
| | Giant silk moths | Increased disease resistance |
| | Greater wax moths | Reduced bruising damage |
| Corn | Wheat | Reduced insect damage |
| | Fireflies | Introduction of marker genes |
| Tomato | Flounders | Reduced freezing damage |
| Tobacco | Chinese hamsters | Increased sterol production |
| Rice | Beans, peas | New storage proteins |
| Sunflower | Brazil nuts | New storage proteins |
| Alfalfa | Bacteria | Oral vaccine against cholera |

Proponents of genetic engineering in agriculture predict that the success of these endeavors has the potential to deliver enormous benefits, particularly to burgeoning populations of the third world that frequently suffer from adverse agricultural conditions. Worldwide, the levels of underground water tables are dropping because of increased use, and there are also concerns that water sources are being contaminated by salinization and pollution. There are hopes that genetic engineering can develop a drought-resistant food crop that will grow in saline water. If protein genes are inserted into corn, consumers will get a more nutritionally complete food. Introducing pest-resistant genes can cut down on the use of expensive and poisonous pesticides. Engineering "antifreeze" genes into a plant can save Florida tomato and orange growers from the disastrous effects of a sudden freeze. There are even hopes of engineering the production of vaccines into plants to produce "edible vaccines" that can be cheaply and easily delivered to remote recipients.

Since 1994, when the first genetically engineered tomato was offered for sale in the United States, many genetically engineered plants have made their way from the laboratory into the field. Other crops that have since been approved include a transgenic virus-resistant squash, insecticidal potatoes, insecticide-resistant soybeans, and pest-resistant cotton. In 1998 it was estimated that half of the cotton fields, 40 percent of the soybean fields, and 20 percent of the cornfields in the United States are planted with genetically altered seeds.

The genetically engineered plants that are now out in the field have been characterized as a "first wave." These plants are designed to resist insects and exposure to powerful weed killers. The next wave of plants is expected to have genes inserted that change their nutritional attributes. Researchers at Tuskegee University, for example, are working to pack more proteins into sweet potatoes, which are a staple in the diet of many children in developing countries. But the process of genetically engineering more complex traits into plants is proving harder and more costly than envisioned. Genetic engineers have discovered that they are limited in the number of genes that can be transferred and the kinds of characteristics that can be modified with existing genes. Such desirable traits as yield, drought resistance, and nitrogen fixation appear to be determined by the interplay of many genes, so they are hard to manipulate genetically. Introducing new genes into plants is also technically difficult, because plants have thick cell walls and are very big and watery, and the nucleus is only a small part.

Among the methods of introducing new genes into plants are natural plant parasites. Chief among the vectors that are used to insert DNA into a plant are various strains of *Agrobacterium*, which invade their hosts via a special tumor-inducing plasmid. Viruses are also used as vectors. The problem with bacterial and viral vectors is that they generally infect only a limited number of plants, the most difficult plants to infect being grasses like wheat and rye, which are important food crops. Another method is electroporation, in which the living material of the plant cell is exposed to foreign DNA and then subjected to pulses of electricity. Still another is to "shoot" the DNA into a target cell using a DNA-coated plastic bullet.

## The Specter of Killer Tomatoes:
## Genetic Engineering and the Public

While agribusiness interests have embraced the promise of genetic engineering with great eagerness, there is broad public distrust of the enterprise. Milk producers who can claim their product is free of RBST—recombinant bovine somatotropin, a genetically engineered hormone that increases milk yield in cows—can charge premium prices. The market for organic foods, even at steeply higher prices, continues to grow. In Western Europe, public distrust of genetic engineering is so strong that several countries have prohibited the importation of genetically engineered agricultural products from the United States. In fact, the genetic engineering of plants and animals is at this point proceeding only a few genes at a time, so many of the wilder public concerns of genetic experiments gone awry remain in the distant future. Moreover, many crops have already been so selected for special characteristics that they cannot survive unless they are taken care of by a farmer. (Special mutant strains of *E. coli*, which are unable to survive outside the laboratory, are used in most genetic engineering procedures.) However, scientists who have called for stringent controls and serious risk analysis on the issue of genetic engineering in agriculture do have legitimate concerns. One of these is that plants that are engineered to resist herbicides so that they can be treated during the growing season will have the effect of increasing herbicide use, a practice to which many ecologists are opposed. Other worries fall under the rubric of "unintended consequences." There are fears that genetically engineered organisms may evolve in unexpected and potentially damaging ways. The first is that a plant with acquired genes that impart various forms of resistance could turn into an invasive weed and thus become difficult to control and eradicate. Another possibility is that the plant could crossbreed with related wild plants, and thus give the wild cousin, which may be a noxious weed, resistance to a particular herbicide. Genetic engineering, it is thought, might compromise the effectiveness of a particular "resistance" gene. Organic farmers often use the *Bacillus thuringiensis* (or BT) bacteria to fight insects because this microorganism produces a narrowly targeted natural insecticide. These farmers fear that the widespread insertion of *Bacillus thuringiensis* genes into crops to combat insect attacks will eventually result in pest resistance to a bacteria that is an important part of their anti-pest repertoire. There are also concerns that the transfer of genes from one species to another could pose a danger to people who suffer from food allergies. Using genes from the Brazil nut, for example, to boost the protein content of a food crop could pose a danger to a person who is allergic to that particular protein but is not aware that the genetic transfer has taken place.

There are also misgivings about the pace at which genetic engineering proceeds. The process of evolution in the natural world is actually quite slow: Advocates for a more cautious approach to genetic engineering quote Werner Arber, a Nobel Prize–winning geneticist at the University of Basel, who has calculated that in the last 4 billion years of life on Earth, there has been time to change and test by selection all the possible combinations of only one hundred bases in each species. Speeding up this process of genetic change, without allowing for millions of years to test their effects, may pose problems that we cannot foresee.

Another concern is that genetic engineering will further compromise the genetic diversity of the world's food crops by favoring the cultivation of the genetically engineered food crop over a more diverse variety of crops. The lack of genetic diversity (or rise of monoculture) was already an issue before the promise of genetic engineering became a reality. Paul Raeburn, an award-winning science writer, has pointed out that the entire U.S. potato crop, worth over $2.5 billion annually, consists of only six varieties, and 90 percent of the sunflower hybrids grown in the United States share the same cytoplasm. Many agriculturists are anxious that this makes the world's food supply dangerously vulnerable if attacked by a newly emerged pest, particularly because there are no chemical substances known that can cure a plant of a virus disease. Worldwide, farmers are locked in perpetual battle with pests, including weeds, insects, viruses, fungi, and bacteria. There is serious worry that in the rapidly mutating world of pests, one will arise that is impervious to the current range of herbicides or insecticides, and devastate an important food crop. In 1970, southern corn leaf blight, caused by a newly emerged strain of the fungus *Helminthosporium maydis*, attacked the American corn crop, a substantial part of which consisted of varieties sharing identical cell cytoplasm. Before the epidemic was over, 15 percent of the crop was lost.

One method of combating plant pests is to find a resistant relative of the food crop, and breed that resistance into the crop. Traditionally the source of these resistance genes has been in the particular food crop's "center for genetic diversity," the part of the world where this particular crop first originated. The "center for genetic diversity" of corn, for example, is Mexico. There are concerns that the introduction of genetically engineered crops into areas of "genetic diversity" could result in the extinction of a wild relative that may have useful genes. Though it is unlikely that genetically engineered corn could have deleterious effects in the United States, where it is not native, it could compromise the genetic stocks of wild corn in its area of origin. These concerns point to one of the arguments for protecting biodiversity, even at a stage where engineers are able to synthesize DNA in the laboratory. At the DNA level, extinction is the loss of unique genes, and we cannot, at present, just invent gene sequences for genetic engineering.

Some proponents of traditional agriculture are also concerned that genetically engineered seeds will further entrench the power of multinational seed companies and thus further impoverish poor farmers around the world, by forcing them to buy expensive seed, fertilizer, and insecticides. A frequently cited example is the 1996 introduction of genetically engineered soybeans that were resistant to Roundup, a herbicide manufactured by Monsanto, the same source of the genetically engineered soybeans. While this enables farmers to kill weeds during the growing season without damaging the valuable crop, it nevertheless potentially locks farmers into using both the genetically engineered soybeans and the company's proprietary—and expensive—herbicide.

## Unintended Consequences:
## The Implications of Genetic Engineering

When physicists split the atom and learned to utilize its energy, observers predicted a glorious future for nuclear weapons and pollution-free energy. This technological breakthrough turned out to be a Pandora's box that many feared could bring about the end of life on Earth in a nuclear catastrophe. Now the box is being repackaged, with test-ban treaties and the dismantling of nuclear weapons. Even the promise of peaceful use of nuclear power faces the enormous and seemingly intractable problem of disposal and storage of nuclear waste.

Now, as they unravel genetic codes and shuffle and manufacture genes, geneticists are homing in on nature's secrets. It may be that the complex chemistry of the living cell will forever elude us, but the implications of deciphering and exploiting the mysteries of DNA are enormous. The beneficial fruits of genetic engineering could be food production that is humane and environmentally benign. We could develop plants that help clean up the environment by absorbing toxic metals or removing carbon dioxide from the atmosphere, a step that could help prevent global warming. There is the promise of fine-tuned medical techniques and the production of complex chemical materials without dangerous by-products. On the other hand, some observers fear that the rash exploitation of burgeoning genetic knowledge, if driven solely by profit considerations and if not utilized with caution, could be enormously destructive.

One worrying element is that genetic research and commerce are intertwined to an unprecedented extent, with many prominent scientists deeply involved in the biotechnology industry. Egged on by the prospect of enormous profits, corporations may rush genetically engineered products to market with too few assessments of their possible long-term effect. This interplay of science and profit has in itself become a minefield of controversy, with some researchers contending that the proprietary nature of so much gene research hinders scientific progress. The heads of biotech companies counter that the financial incentives of patenting genetic sequences and research are necessary to help fund the expenses of these investigations. "We are busily developing a science, and a resultant group of high technologies, that could change the world and our attitude to it more radically than any we have seen before," writes Colin Tudge, an eminent British science writer. "If this science and these technologies are deployed adroitly, they could do more than any other sciences or techniques to solve our present problems and avert the pending disasters of the future . . . [but] if we fail to remain in control, then the science and new technologies could equally be employed for ends which we may well consider evil, or at least deeply sinister, and which, if things go wrong, could trigger a chain of biological destruction that could outstrip any we have seen before."

---

## A BIOTECHNOLOGY TIMELINE

Rapid developments in biotechnology have followed close on the heels of basic discoveries in genetics. The following is a chronology of some of the most important events.

1973    Stanley Cohen and Herbert Boyer invent recombinant DNA, a method of merging the genetic material of different species.

1975    Edward Southern develops a method known as Southern blotting, to pinpoint a specific genetic sequence. Howard Temin and David Baltimore receive the Nobel Prize for the discovery of reverse transcriptase. This enzyme is encoded by retroviruses, and converts RNA into DNA.

1977    The production of the first human protein manufactured in a bacteria is reported. The insulin gene in rats is isolated. Walter Gilbert and Allan Maxam at Harvard University, and Fred Sanger, working at Cambridge, devise methods for rapidly determining the sequence of bases in long sections of DNA.

1978    Human insulin is successfully produced using recombinant DNA technology. Researchers discover variations in DNA from different individuals when a restriction enzyme is applied. The DNA variations are called *restriction fragment length polymorphisms*, or RFLPs, and they will prove to be extremely useful in genetic studies

1979    Researchers identify the gene for a protein in the virus that causes hepatitis B. The gene for human growth hormone is cloned, or identified.

1980    The U.S. Supreme Court rules that genetically altered life forms can be patented, thereby opening up enormous possibilities for commercially exploiting genetic engineering. Kary Mullis and others at Cetus Corporation in Berkeley, California, invent a technique for multiplying DNA sequences using the polymerase chain reaction (PCR).

1981    Scientists produce the first transgenic animals by transferring genes from other animals into mice. The gene for insulin is mapped, and a gene-mapping or identifying method, called *in situ hybridization*, becomes standard.

1982    Genentech, Inc., receives approval from the U.S. Food and Drug Administration to market genetically engineered human insulin. An agricultural technology company requests government permission to test genetically engineered bacteria that can control frost damage to potatoes and strawberries.

1983    The AIDS virus is isolated in the United States and at the Pasteur Institute in France. Patents are granted in the United States for genetically engineered plants. A study of an extended family in Venezuela with Huntington's chorea demonstrates that family members with the disease show a distinct and characteristic pattern of restriction fragment lengths. The same methods of investigation reveal patterns for cystic fibrosis, adult polycystic kidney disease, Duchenne muscular dystrophy, and others. Researchers develop a method of making DNA fragments of predetermined sequence automatically.

1984    A gene that helps regulate blood pressure and control salt and water excretion is isolated. The entire human immunodeficiency virus (HIV) genome is sequenced and cloned.

1985    The gene that encodes human lung surfactant protein is cloned, a major step toward reducing a premature birth complication. PCR technology, which can generate millions of copies of a targeted gene sequence in hours, is published. Genetically engineered plants resistant to insects, viruses, and bacteria are field tested for the first time. The NIH approves guidelines for performing experiments in gene therapy on humans.

1986    Scientists and technicians at the California Institute of Technology and Applied Biosystems, Inc., invent the automated DNA fluorescence sequencer. The FDA grants a license for the first recombinant vaccine (for hepatitis) to Chiron Corp. The EPA approves the release of the first genetically engineered crop, gene-altered tobacco plants. The first disease gene detected by positional cloning is identified, that for an immune disorder called chronic granulomatous disease.

1987    The FDA approves the marketing of a genetically engineered tissue plasminogen activator that is used to treat heart attacks. Calgene, Inc., receives a patent for the tomato polygalacturonase DNA sequence, used to produce an antisense RNA sequence that can extend the shelf life of fruit. Advanced Genetic Sciences, Inc., conducts a field trial of a recombinant organism, a frost inhibitor, on a Contra Costa County strawberry patch. Maynard Olson and colleagues at Washington University invent "yeast artificial chromosomes," or YACs. This is a method of incorporating the gene for a large protein into a yeast, where the protein can be manufactured.

1988    Philip Leder and Timothy Stewart, molecular geneticists at Harvard, introduce the "Harvard Mouse"—a line of genetically engineered laboratory mice, and receive the first patent for a mammal. An immune-deficient mouse with a reconstituted human immune system that has been engineered for AIDS research is patented, as is a process for making bleach-resistant protease enzymes to use in detergents.

1989    UC Davis scientists develop a recombinant vaccine against the deadly rinderpest virus, which had wiped out millions of cattle in developing countries.

1990    The first successful field trial of genetically engineered cotton plants is conducted. The plants had been engineered to withstand use of the herbicide Bromoxynil. The engineering of genes in corn is reported. Epidemiologists report the discovery of the gene linked to early-onset breast cancer in families. The first transgenic dairy cow is created. The cow is used to produce human milk proteins for infant formula. A four-year-old girl suffering from ADA deficiency, an inherited disorder that destroys the immune system, becomes the first human recipient of gene therapy. The Human Genome Project, the international effort to map all of the genes in the human body, is launched.

1991    The celebrated reference work *Mendelian Inheritance in Man* is made available through an on-line computer network. The catalog lists some 5,600 genes known or thought on good evidence to be inherited in Mendelian patterns.

1993 Kary Mullis wins the Nobel Prize in Chemistry for inventing the technology of polymerase chain reaction (PCR). Researchers discover that defects in a gene that corrects genetic errors are responsible for a hereditary form of colorectal cancer.

1994 The BRCA I gene is "discovered." Mutations in this gene are linked to hereditary breast cancer.

1995 Monsanto introduces a genetically engineered potato that is the first commercial crop to be protected from an insect pest through biotechnology. The Institute for Genome Research completely decodes the genome of a bacterium, H. influenzae, following shortly afterward with the complete genome of a mycoplasma.

1996 Full sequencing of the BRCA II gene is reported. The complete genomes of a deep-sea bacterium and that of a species of yeast are spelled out.

1997 Scientists at the Roslin Institute announce the birth of Dolly, a Dorset sheep that they claim is the first animal cloned from an adult cell. Monsanto introduces cotton, soybeans, and corn that are genetically engineered to resist the Roundup brand of herbicide. The Japanese government approves construction of a facility dedicated to sequencing the entire mouse genome. The Institute of Genome Research completes sequencing the genome of Borrelia burgdorferi, which causes Lyme disease.

1998 An international team at the University of Hawaii clones several generations of mice using a different method from that used by Scottish researchers to clone a sheep. The genome of a microbe, Dernococcus radiocurans, which can withstand massive amounts of radiation, is completed. The genome of the first animal, a microscopic roundworm called Caenorhabditis elegans, is completed.

# 6

## Medicine and Molecules: New Approaches to Drug Development

Advances in molecular biology—the study of molecules like DNA, RNA, and the proteins and sugars that are the basis of life—are changing the focus of much medical research. The investigation of diseases has become increasingly less a study of organs, tissues, and cells than an exploration of molecular processes. Microbiologists and molecular biologists are rapidly gaining new insights into the complicated chain of chemical reactions, known as biochemical pathways, that the life processes of all living organisms follow. With these insights, like identifying the specific molecular structure of enzymes that govern the cascade of chemical reactions that relay messages between cells in a particular tissue, for instance, researchers are devising new methods of combating viruses or diseases of the immune system, like arthritis. Similarly, the ongoing intensive research into the human genome, which is unraveling the genetic code contained in human DNA, is swiftly revealing the function of specific genes and the consequences of their malfunction. As genes implicated in diseases are discovered and their DNA sequences are revealed, new targets and approaches for disease treatment are suggested.

These advances in molecular biology are expected to revolutionize the kinds of drugs and treatment doctors will prescribe for an ailment.

Many drugs and medical treatments lack precision. Most drugs cause side effects. Implanted devices cause inflammation and rejection. Current cancer treatments can often be as damaging to the patient as the illness. The emergence of resistant bacteria signals an urgent need to develop antibiotics with novel mechanisms of action. As molecular biology research reveals disease targets of molecular dimensions, observers predict the development of a new generation of medical treatments and drugs that will reach their targets with much more accuracy and refinement.

The quest for precisely targeted drugs is reflected in current pharmaceutical research. The multibillion-dollar international drug industry is actually dependent on relatively few compounds, and the hunt for new drugs is relentless and expensive. Raymond V. Gilmartin, chief executive of Merck & Co., Inc., the giant pharmaceutical company, has noted that only one compound out of five thousand new discoveries ever makes it to the marketplace. "And getting there," he said, "costs over $400 million, and takes fifteen years of testing and regulatory review."

The traditional search for effective drugs takes various approaches, one of which involves seeking out compounds synthesized by living organisms like plants, fungi, and bacteria to fight their own enemies. This process requires screening tens of thousands of compounds from sources as diverse as garden soil and coral reefs. Once a compound has been identified and isolated, pharmaceutical chemists devise methods of obtaining useful quantities from natural sources, or else of synthesizing it or a similar compound that is even more effective. For example, cyclosporine, a drug that is an indispensable adjunct of organ transplants because it suppresses rejection, was first garnered from a bacterium found in the soil of the Norwegian tundra.

This previously somewhat random—and very costly—method of drug discovery is being supplanted by a more focused quest for compounds that can facilitate or prevent a biochemical reaction. This is possible because molecular biology research is revealing many more potential drug targets, including their chemical composition and three-dimensional structure. For example, knowledge of the structure and function of an enzyme that triggers an inflammatory response in the body enables investigators to develop a drug that is shaped to combine precisely with that enzyme and thus block its action. Researchers are also studying the methods that bacteria and viruses use to invade human cells. Molecular insight into the various survival mechanisms of bacteria is enabling pharmaceutical chemists to experiment with drugs that attack different targets from those of traditional antibiotics. One approach would be to block bacterial kinases, which are the chemicals that bacteria use to respond to their environment. Blocking the chemical portal that a virus uses to gain access to a cell can prevent infection. This method of developing a drug that will interact in a specific way with a targeted molecule is known as *rational drug design*.

Rational drug design is one strategy being used to develop drugs to combat the human immunodeficiency retrovirus that causes AIDS. Protease inhibitors, the drugs that are currently enjoying the most success in treating AIDS, were successfully developed once the structure of the HIV protease enzyme was published in 1989. (HIV protease is an enzyme that mediates the transformation of immature HIV into the mature form of the virus.) Another avenue of research is to develop another kind of molecule that will bind tightly to the portal used by the virus to invade lymphocytes and other types of blood cells. These members of the body's immune system are called chemokines. They are anti-inflammatory agents whose natural function is to attach to white blood cells, which are the body's major disease fighters, and guide those cells to the site of an infection. Research has shown that chemokines attach themselves to white blood cell surfaces at the same points, called *receptors*, that are used by HIV, thus blocking its access to the cell. The task is to develop chemokines that are capable of blocking various strains of HIV so that they can counteract multiple forms of this rapidly mutating retrovirus. If this is successful, researchers believe that periodic injections of these chemokines could create a barrier between HIV and its target cells, and prevent the virus from spreading its deadly infection.

## Sense and Antisense:
## Crafting the Means to Block Defective Proteins

A related approach to drug design that may hold much promise is based on the burgeoning knowledge of the DNA codes that make up the human genome. Among DNA-based approaches to drug development are the use of recombinant DNA technology to produce such previously scarce pharmaceuticals as interferon and the crafting of agents that can correct, in some way, defects in genes that are the cause of disease. Vaccines consisting of a small amount of DNA from the infectious agent are proving to be an effective means of imparting protection from diseases with just a squirt of nasal spray, and are being explored for more applications (see page 135). Another tactic is called antisense technology. This is based on the knowledge that DNA directs the formation of RNA, which then in turn travels from the nucleus to structures in cells called ribosomes, where protein is manufactured. Some diseases are caused when defective genes give rise to incorrect RNA templates, which in turn lead ribosomes to make abnormal proteins. Antisense agents attempt to compensate for the

genetic damage by binding to the faulty RNA and thus blocking the manufacture of the errant protein. This is accomplished by developing short
nucleotide sequences—called *antisense nucleotides*—that prevent a particular protein from being manufactured. They do this by binding with a
crucial sector of the RNA sequence that would have carried the message
from the DNA to the ribosome. Though still mostly experimental, the
antisense concept could fulfill the pharmaceutical chemist's dream of a
totally selective drug, and it could revolutionize future drug discovery.

Simple as the antisense concept may appear, its implementation is
quite complex, and medications consisting of strings of nucleotides face
many hurdles before they routinely reach the shelves of your local phar-
macy. Researchers have successfully overcome the first obstacle: they have
identified target genes and their nucleotide sequences as well as optimal
smaller target sites within a gene sequence that appear to successfully
block its function if bound by antisense nucleotides. The next steps are far
more problematic, and are typical of the kinds of barriers that drug developers must overcome to develop a new medication. The first big problem
is getting the antisense nucleotide into the cell without its being destroyed
by nucleases (enzymes that would cut it up), which can be found in the tissues around the target site. There is the additional danger that this degradation could even make the nucleotide toxic. To block the action of
nucleases, researchers are experimenting with enclosing the antisense nucleotide in liposomes, which are small synthetic spheres that are hollow
inside (see chapter 8). They have also experimented with slightly changing the chemical composition of the nucleotides to make them more stable. Another challenge is to modify the antisense drug so that it penetrates
the cell wall. Then there is the possibility that nucleases within the cell
will destroy the antisense compound before it binds with the target RNA.
Although an antisense nucleotide and its target RNA join together readily in a test tube, this environment is somewhat different from that inside
a cell; thus the antisense nucleotide will probably have to be modified to
facilitate its linkage with a target sequence of RNA under cellular conditions. Moreover, researchers have to accomplish all of these goals without
compromising the basic function of the antisense compound.

Although many other issues have still to be resolved, researchers
have already successfully introduced antisense compounds into the lung
tissue of experimental animals. Because lung tissue has a large surface area
and nucleotides can easily be administered by aerosol, the lung has become one of the first sites where the potential of antisense can be tested.
More recently, an antisense treatment for cytomegaloviris retinitis, an
HIV-related opportunistic infection that can result in blindness, has been
successfully injected into the eye. In August of 1998, this antisense

oligonucleotide received FDA approval. It became the first antisense drug to receive this imprimatur.

## Compounds by the Million: Combinatorial Chemistry Speeds Up Drug Discovery

In concert with rational drug design, drug companies are using a process called *combinatorial chemistry* to make large numbers of compounds that may turn out to be new drugs. This approach is also used to develop useful new chemical compounds in other fields (see chapter 8). Because chemical elements can be linked to each other in countless different ways, combinatorial chemists use varying systematic approaches, such as applying different amounts of heat or pressure, or varying proportions of chemicals, to quickly produce millions of compounds. The biological activity of these compounds is then tested, or assayed, often using robots that insert the compounds into banks of miniature test tubes containing cells or organic chemicals, or dab small amounts of a chemical onto a polystyrene bead containing the substrate. Optical scanners can then be used to assess whether the compound and its target have reacted together. This process is known as *high-throughput screening*. The results of assays are used to garner information about the relative potency of different molecules. These different but related molecules are designated a *combinatorial library*. Because of the sheer volume of information that this kind of research generates, computers are used extensively to analyze results and develop data about particular molecules. This rapid production of compounds allows researchers to identify candidates for clinical trials faster and at lower cost than before.

This kind of data manipulation is a vital part of a new computer-related specialty called bio-informatics, a discipline that is used to help make sense of the enormous volumes of information generated by combinatorial and genomic research. Among computational approaches to bio-informatics is the development of programs that come up with promising drug compounds without even going near a laboratory bench. One company uses genetic algorithms, a combination of artificial intelligence and fuzzy logic, to have computer software develop the most promising chemical candidates for a drug (see chapter 2). The computer program starts off with a molecule known to provide a useful activity. The software produces a number of variations by randomly rearranging the atoms and substituting some atoms for others. Each molecule is evaluated for such properties as toxicity and stability, and the compound with the highest score then becomes the "parent" of more variations. The software runs this process through thousands of generations. Its developers say the process has produced promising drug candidates, particularly prospects for new, small-molecule antibiotics.

## Biotechnology and Cancer: The Search for New Weapons

Probably the most intensive molecular focus in recent decades has been on the nature and treatment of cancer. The amount of information gathered over the past twenty years about the genetic and molecular origins of cancer is without parallel in the history of biomedical research. Biotechnology experts believe that we are close to tracking the life histories of many human cancers from start to finish, and that within the next ten years we will understand precisely how a normal cell evolves into a highly malignant and invasive tumor. The growing body of knowledge is leading to a range of new strategies to combat cancer. Though it may be years before these new approaches bear fruit, cancer specialists are more optimistic than they have been in decades that long-awaited breakthroughs in diagnosis and treatment are on the horizon.

The developments that will probably make their way into medical practice first are new methods of diagnosing and assessing the stage that a malignant tumor has reached. The ability to diagnose cancers accurately, especially before they have metastasized, or spread to other parts of the body, is medically advantageous. Cancers are more easily cured in the early stages. With accurate staging, a patient can get the most appropriate therapy. If a diagnosis reveals advanced cancer, more aggressive treatment is warranted.

The consensus that is guiding new diagnostic strategies is that cancer cells in a tumor descend from just one cell whose genetic machinery was altered so that it began a program of uncontrolled reproduction. This process usually begins many years before a tumor can be detected, and it is the result of mutations in at least six—or many more—of the genes in the cell that direct its life cycle. Although, at this time, the discovery process is far from complete, the developing genetic and molecular understanding of cancers has enabled researchers to build molecular tools for detecting and determining the aggressiveness of the disease. This new range of diagnostic tests, which look for telltale alterations in certain key genes, will save lives because they can aid early detection of cancers when treatment is most likely to be successful, and they will be able to diagnose the stage of the disease accurately. They are also expected to be useful for assessing whether a person is at risk for developing a certain kind of cancer, and for discovering tumors or cancer cells circulating in a person's body.

Other strategies for early detection have focused on monitoring the levels of proteins that are either the product of a mutated gene or are present as a result of the unique biochemistry of a particular cancer. For example, high levels of a protein known as *circulating PSA* (prostate-specific antigen) are found in the blood of patients with prostate cancer. Supporters of the PSA test contend that it has saved a significant number of lives because it enables early diagnosis of prostate cancer. Another, more experimental, protein test looks for telomerase, an enzyme that is active when cancer arises. This enzyme blocks the shortening of the *telomeres*, which are the segments at the end of chromosomes that grow progressively shorter each time a cell divides. In normal cells, when telomeres shorten to a certain length, they instruct the cell to self-destruct, providing a mechanism to rid the body of aging cells. The presence of large amounts of telomerase in a cell indicates that its regular life cycle has been disrupted.

The development of ultrasensitive tests has enabled researchers to look for revealing changes in cell chemistry. These include such techniques as the polymerase chain reaction to culture sufficient quantities of DNA for genetic analysis, the development of "gene chips" that can be used to diagnose genetic abnormalities, and monoclonal antibody tagging to identify the presence of a particular receptor on a malignant cell (see page 161). New tests are expected to be helpful in determining the course of treatment. These include the ability to detect low levels of cancer cells, called *micrometastases*, in the bloodstream. Another test can identify markers that distinguish a cell as having come from a specific tissue or type of tumor.

Some observers predict that within the next few years, molecular detection—and subsequent strategies tailoring treatment approaches—will be part of a routine physical examination for most people in developed countries. By analyzing genes or detecting the presence of certain proteins, physicians will be able to determine from just a drop of blood or urine whether a person is at special risk for a cancer, or has an unnoticed microscopic tumor instead of using invasive probes or lengthy scans that can pose risks of radiation. For people at risk, prevention strategies—such as changing behavior and diet, or taking medications—may be available. For those who already have cancer, analysis of the tumor's genes will reveal how aggressive it is, whether it needs extensive treatment, and which therapies might be effective.

## Sharpening the Blunt Instrument: Molecular Strategies against Cancer

The traditional weapons in the war against cancer are blunt instruments with side-effects that, at their worst, can threaten the lives of cancer patients. Depending on the kind of cancer, the first line of defense is surgery to remove the primary tumor. Then, to kill malignant cells that may have made their way into the bloodstream, cancer specialists prescribe a course of chemotherapy. Many chemotherapeutic agents kill cancer cells by blocking their growth. However, drugs that kill a rapidly dividing malignant cell will also destroy rapidly dividing healthy cells, such as those in the intestinal lining, hair follicles, or blood. They may also cause long-term damage to such vital organs as the heart or liver. Thus a treatment that shrinks the tumor may also cause vomiting, hair loss, anemia, and a dangerous drop in white blood cells that can put a patient at risk for infection. Chemotherapy is usually successful for a while, but some cancer cells might survive this attack. In a process that is analogous to the phenomenon of antibiotic resistance, the surviving cancer cells are generally resistant to another round of chemotherapy and are frequently the most malignant. If they go on to form secondary tumors, the cancer is likely to eventually prove lethal.

Although chemotherapy research continues, and it is likely to be the "state of the art" for the near future, many researchers now believe that indiscriminately blasting the body with chemotherapy is an outdated approach. Armed with a greater understanding of the genetic and molecular pathways of cancer, researchers are now focusing on key abnormalities that distinguish a malignancy from "normal" cells. They then hope to develop drugs that attack only malignant cells. This approach, they hope, will yield new drugs that act more like guided missiles and less like carpet-bombs. Strategies that attack cancer at the molecular level include efforts to repair the genetic damage that gave rise to the malignancy, shutting down important growth proteins so that the cancer cell will die, and increasing the sensitivity of a tumor cell to conventional therapies like radiation.

Based on current research, specialists believe that cancer results from mutations in classes of genes that are responsible in some way for the reproduction of cells. The defects occur in three types of genes: *oncogenes*, which stimulate the cell to reproduce; *tumor suppressor genes*, which restrict the uncontrolled growth of the cell; and genes that govern the replication and repair of DNA. These mutations alter the proteins that growth-regulating genes code for, and in so doing disrupt "normal" controls on cell division. This theory has been supported by studies of tumor cells in cancer patients, which have revealed mutations in these classes of genes in many cases. Indeed, some researchers predict that in coming decades, cancers will not be described by the organ they attack, like the lungs, but by the genes whose functioning has gone awry. Thus, one area of pharmaceutical research is to develop new drugs that will act on these disrupted genes or their proteins. It is hoped that these drugs can either return malignant cells to normalcy, or kill the cells without harming their healthy neighbors.

The difference between oncogenes and normal genes can be so subtle that the mutant protein an oncogene creates may differ from the healthy version by one amino acid. But that one alteration can radically change the protein's function (and has proved fatal in other diseases of genetic origin, such as Duchenne's muscular dystrophy). Defective *Ras* genes, for example, are found in several human cancers. When damaged or mutated, this gene directs the cell's machinery to manufacture a defective protein that then triggers aggressive cell division. To counter this, researchers seek out a vulnerable spot in the biochemical processes initiated by the gene. In the case of the *Ras* gene, one target that researchers have homed in on is a cell enzyme called *farnesyl transferase*. If this enzyme is inactivated, the mutant *Ras* protein can't instruct a cell to divide. Investigators are working on determining the three-dimensional structure of the enzyme, so that they can tailor a drug that fits snugly onto the site of the enzyme that is active in facilitating the protein's action. Antisense therapy is also being investigated as a promising tool to disable malfunctioning genes. This tactic blocks the production of damaging proteins by binding to important sequences on the messenger RNA that carry the instructions encoded in the gene to the ribosomes where the protein is assembled (see page 113). Researchers have, for example, reported encouraging results in animal experiments using antisense oligonucleotides together with other drugs to kill melanoma cells.

Another focus is on mutations in a tumor suppressor gene called P53 that, when it is working properly, is responsible for keeping the fragile DNA of the cell in order as it reproduces. In normal cells, the protein encoded by P53 prevents the replication of damaged DNA and promotes cell death, or apoptosis, if the DNA is irreparably damaged. In genetic studies of many human cancers, defects in the P53 gene have appeared. To counter this, researchers are attempting to introduce normal P53 genes into tumors. One approach uses an adenovirus that would usually cause a mild cold to carry the gene into the cell. In this experimental treatment, the virus is genetically altered so that it only becomes active in cells with the defective P53 gene. This approach has reportedly had some success and is being investigated further.

Other anticancer targets include genes that encode enzymes called *protein kinases*. In normal cells, protein kinases help to regulate many important processes, such as sending signals between the cell membrane and the nucleus. Researchers are exploring the possibility of identifying kinases that are only found in cancer cells and then developing a method of blocking their action and thus, possibly, killing the cell. The problem, which illustrates the enormous complexity of microbiology and cancer research, is that there are nearly one thousand protein kinases in mammalian cells! Another option may be to develop methods of interfering with cancer cell adhesion. A normal cell recognizes its proper place in the body by fitting its adhesion molecules to those on other cells and on the extracellular matrix. In cancer, something goes wrong with the address system. Cancer cells appear to be capable of anchoring themselves to a variety of tissues. Researchers are working on identifying the molecules in the cancer cell that permit it to adhere to foreign tissues. Once those substances are identified, the researchers will develop methods of blocking their action. Yet another target is telomerase, the enzyme that restores the telomere. Several pharmaceutical firms are attempting to develop drugs that block the functioning of telomerase and thus end the cell's reproduction cycle.

## Making Molecular Blockades: Preventing Angiogenesis

Another approach is to block the tumor's access to nutrients and oxygen by preventing the development of blood vessels in the tumor. This blood vessel growth, called *angiogenesis*, enables the malignant cells to continue dividing. Many researchers believe that tumor cells emit chemical signals, called *growth factors*, toward the nearest blood vessel. In response, the vessel then sends a new branch toward the site of the signal. In recent years, scientists have identified at least fourteen different proteins found in the body that trigger blood vessel growth. Blocking the tumor's protein message, researchers say, could deprive the tumor of its blood supply. One approach is to stop the chemical signals from binding with the surface of the blood vessel cells. At present, more than one hundred different substances are being tested for the potential to block angiogenesis. These include thalidomide, a tranquilizer that caused many birth defects in the 1950s because it cut off the blood supply of developing fetuses. Researchers are trying to produce drugs that dock with these blood-cell receptors and prevent the tumor's signal from reaching its destination. Another strategy is to use genetic engineering. Investigators have, for example, inserted the gene for platelet factor 4, a substance that stops the growth of blood vessels, into an adenovirus. The adenovirus serves as a vehicle, or vector, to deliver the gene into the target cells. When the altered adenovirus was injected into the tumor, researchers found that blood vessel proliferation appeared to be inhibited. In 1998, researchers reported the isolation of two angiogenesis inhibitors, called *endostatin* and *angiostatin*, that were remarkably effective in eradicating large tumors in mice. In their experiments, the researchers obtained minute quantities of these substances from mouse urine. Efforts are under way to synthesize or extract them from other sources. But until larger quantities of the substances are manufactured, crucial trials of these substances on human subjects cannot be done. Other angiogenesis inhibitors, however, are already in a later stage of clinical studies needed for approval by the U.S. Food and Drug Administration.

## The Light That Heals: Photodynamic Therapy

Photodynamic therapy (PDT) is a promising approach to cancer treatment that combines the precision of a laser beam with drugs that only become active when they are exposed to a specific wavelength of light (see chapter 4). It takes advantage of the fact that these kinds of drugs accumulate in larger concentrations in tumors than in normal tissue. PDT has already been approved by the U.S. Food and Drug Administration to treat tumors that obstruct the esophagus, and other uses are likely to be approved in the near future.

Patients with advanced cancer receiving photodynamic therapy are given drugs that are preferentially taken up in tumor cells. After the drug is administered, doctors using lasers shine focused light on the tumors. This light energy is captured by the drug and is then passed on to nearby oxygen molecules. These activated oxygen molecules, which react with almost any molecule in their immediate vicinity, inflict fatal damage on the cancer cells. After the light is removed, however, the chemicals become harmless within minutes, and patients are able to walk away from the treatment table. They must, however, limit their exposure to the sun, because the chemicals used in PDT make them sun-sensitive for several weeks after treatment.

Cancer researchers have been attempting PDT with mixed success since the 1970s. Early research foundered because investigators lacked photosensitizing drugs with tolerable side-effects. In recent years, however, there has been an explosion of new drugs with fewer side-effects. In addition, these new drugs are activated at higher—and thus more energetic—wavelengths of light. This allows light to penetrate tissues more deeply, and thus kill tumors more effectively. In some cases, doctors use endoscopes to deliver laser light to the tumor site. The combination of new drugs with improvements in laser technology is increasing the potential range of this new therapy.

In Europe, PDT is increasingly being used to treat skin cancers and to help detect and delineate skin tumors with badly defined borders. Patients are treated with a photoreactive chemical; their skin is then inspected under an ultraviolet light. In Japan, PDT has been successfully used as an alternative to surgery to treat early cases of lung cancer. Although the cure rate for the Japanese patients is similar to that with conventional treatments, PDT is preferable to the removal of a vital body organ. It is also a more cost-effective and simple therapy. Among other possible targets for PDT are tumors of the brain, head, neck, breast, esophagus, pleural cavity, abdominal cavity, bladder, prostate, and cervix. PDT is also being explored to treat conditions such as endometriosis and psoriasis.

Even while this promising therapy is in its experimental stages, new materials are being developed that can overcome some of its current limitations. The light-sensitive drugs presently used as PDT drugs are activated only by visible light, and work only in the presence of oxygen. Thus they can only destroy thin and flat tumors, or malignancies in early stages of development. New materials that are being developed at the Weitzmann Institute in Rehovot, Israel, are derived from chlorophyll, the green pigment that allows plants to capture sunlight. They can be applied to larger solid tumors because these materials absorb both visible and near-infrared light that penetrates much deeper into body tissues. Green materials can also work with low oxygen levels, a condition that often occurs inside large tumors. Another advantage of the new chlorophyll derivatives is that they clear out of the body much faster than existing drugs. Thus patients can tolerate outdoor light shortly after treatment without fearing that the photosensitive materials will harm their skin.

To deliver their materials precisely to the desired site, the Weitzmann Institute scientists have developed a method of attaching them to antibodies or other molecules that act as guided missiles, transporting the materials to the target in the body.

## Lethal Weapons: Nuclear Medicine Attacks Cancer

The Weitzmann Institute scientists, who hope to deliver chlorophyll-derived photoreactive chemicals to cancer tumors using antibodies, are among many researchers who are developing vehicles that will deliver a lethal punch to cancer cells while leaving healthy cells untouched. Monoclonal antibodies are a focus of interest among many researchers.

Because monoclonal antibodies form in response to a specific antigen or invader, scientists are experimenting with a new nuclear therapy that combines a specifically tailored antibody with a radioisotope that destroys the cancer cell when the antibody finds its target. This therapy would minimize harm to normal tissue both because the antibody is tailored to a cancer-specific antigen and because the isotope used emits particles that can only travel short distances in tissue.

## MAGIC BULLETS

Antibodies are Y-shaped proteins produced by the body's white blood cells (or lymphocytes) to fight foreign substances, or antigens. The antibody's molecules are shaped so that they fit, like a lock into a particular keyhole, into the molecules on the surface of a particular antigen. This specificity enables the antibody to recognize and destroy a particular foe. The body's immune system develops antibodies in response to an encounter with an invader, like the virus that causes chicken pox. The bloodstream of an adult animal can contain millions of kinds of antibodies designed to home in on and destroy a particular invader. Developing techniques to produce a specific antibody in large quantities is considered one of the most important accomplishments in biotechnology in recent decades.

In the 1970s, scientists found a high concentration of one type of antibody in the blood of people with a malignant tumor called multiple myeloma. They devised a method of fusing these myeloma cells with lymphocytes from tissues of mice that had been exposed to an antigen. The resulting hybrid cells were able to produce large amounts of a specific antibody. These are called *monoclonal antibodies*. Using similar principles, scientists have developed methods of obtaining an antibody that combines with any chosen foreign substance. Monoclonal antibodies have become valuable tools in biology and medicine because these antibodies can combine with and identify the component substances of cells and tissues, or can be used to target drugs to specific sites in the body.

Monoclonal antibodies were dubbed "magic bullets" when they were first invented in the 1980s, because it was widely expected that their ability to home in on a particular target in the body would make them the ideal method of curing many diseases. In fact, they did not initially live up to their promise, mainly because the body recognized and rejected the mouse elements of the antibody. Researchers have since developed methods of making monoclonal antibodies more "human," so that they are not rejected by the immune system, and they are coming back into favor. In September 1998, the FDA approved the first monoclonal antibody drug for cancer. The drug, Herceptin, blocks overactive receptors in the malignant breast cancer cells of patients who have a particular form of the disease.

Current research is also aimed at producing monoclonal antibodies more economically. Scientists have successfully harvested monoclonal antibodies from genetically engineered plants.

As in other proposed therapies, such as antisense (see above), though the method appears ingenious, its implementation is complicated and arduous. A radioisotope is a single atom of an unstable element, while an antibody is a Y-shaped protein. To bring both together chemically, biologists and chemists have had to engineer molecular cages to hold the isotope, and linking molecules to bond the cage with the antibody. It took ten years to construct the "cage" for one species of isotope, and so far less than a handful of isotopes and antibodies have been successfully combined.

Monoclonal antibodies are also being used in experimental therapies to exploit the immune system to stop the development of cancer. In research on developing cancer vaccines, researchers are attempting to use monoclonal antibodies that will home in on cancer cells to trigger an immune attack that destroys the targeted cells. However, although monoclonal antibodies have revealed a large array of antigens that exist on human cancer cells, researchers have found that most of these antigens are also found on normal cells. Antibody engineers are working on identifying cancer-specific antigens and on other aspects of the antibody molecules to make them better able to bind to antigens and penetrate tumors.

Monoclonal antibodies are one approach to stimulating the immune system to attack and kill cancer cells. Another strategy is to remove some of a patient's cancer cells and, in the laboratory, to modify them with genes encoding for cytokines. These are immune system proteins that act as messengers, and it is hoped that inserted cytokines would alert the immune system to its malignant invader. One problem with the vaccine approach is that each cancer would have to receive individualized treatment, an option that may be too costly to be broadly adopted.

### The Long and Winding Road: Many Steps to Drug Approval

There are hopeful signs that ongoing research into new and more effective treatments for cancer will result in the development of an arsenal of therapies for a range of cancers within the next ten years. Investigators predict that this new generation of drugs will have few side-effects, and can be taken in large doses for many years. But for metastatic cancer to attain the status of a chronic but manageable disease, the new medicines based on advances in molecular biology will still have to overcome the same kinds of obstacles any new drug must overcome. The drugs must be able to locate their target and they must be structured so that they are stable and can get into the malignant cells in sufficient quantity to be effective. Because pills are easy to take, pharmaceutical chemists strive to package new drugs in this form. This rules out proteins, for example, because they are large, fragile, and would be digested if taken in pill form, although they can be injected. They also cannot cross the blood-brain barrier. Thus, researchers are intensively studying the potential of a piece of a protein, called a *peptide*, that is a key portion of a larger protein. The drug itself must be tested for toxicity and side effects, as well as evade the same kinds of problems of resistance encountered by traditional chemotherapy.

Drugs that have proved potent in animal studies would have to be carefully assessed in human trials. There have been many instances in which drugs that cured cancer in animals have not had the same effects on humans. This caution was sounded in response to the euphoria that greeted the news, in 1998, that two angiogenesis inhibitors, angiostatin and endostatin, had apparently eliminated cancerous tumors in mice. Dr. Judah Folkman, who developed the approach, cautioned a reporter for the *New York Times*, "If you have cancer, and you are a mouse, we can take good care of you."

Among the steps in the process are two to three years of molecular, genetic, and cell biological studies to confirm that an identified target is critical to the development of human cancers. This is followed by about two years of biochemical screening assays, or tests, to find promising compounds. Once a good possibility is discovered, medicinal chemists modify the drug to optimize its potency, specificity, and other pharmacological properties. This usually takes three to five years and requires the synthesis of several hundred to several thousand related compounds. This process is followed by animal testing, and then clinical trials in which the safety, efficacy, and proper doses of a drug are determined. Thus, experts believe that even using the latest methods of pharmaceutical science, new anticancer agents currently in the pipeline will take several more years to become available and cost hundreds of millions of dollars to develop.

# 7

# *Spare Parts and High-Tech Flashlights: Repair Kits and Diagnostics for the Human Body*

Since the turn of the century, modern medicine has produced a plethora of such life-saving devices as the heart-lung machine, kidney dialysis, and cardiac pacemakers. Today's high-technology medicine continues to take its inspiration and tools from the leading edges of many disciplines, including microelectronics and computer science, materials science, and particle physics. Emerging from this storehouse of knowledge are medical specialties like endoscopy, telemedicine, and tissue engineering. Indeed, the same technologies that have brought powerful personal computers, inexpensive long-distance telephone service, and the Internet into the homes of numerous middle-class Americans have enabled the development of lifelike artificial limbs and minimally invasive surgery using fiberoptics. Throughout the trillion-dollar enterprise that is modern medicine are numerous examples of procedures and devices that exploit human ingenuity and knowledge from a broad range of disciplines.

## Medical Imaging and Microchips:
## The Journey from X Rays to 4D Endoscopy

One field that has tapped expertise ranging from physics and microelectronics to molecular biology is medical imaging. Just over a century ago, Wilhelm Conrad Roentgen, passing a high-voltage electrical current through a glass tube with most of its air pumped out, made an X-ray picture of his wife's left hand. The fifteen-minute exposure of Frau Roentgen's finger bones used X rays, high-frequency electromagnetic radiation, that could pass through opaque objects and leave an impression on a photographic plate. Over the succeeding century, medical imaging has progressed remarkably in its depth and capabilities. New X-ray machines do not use film or photographic plates. Instead, X-ray images are digitized and displayed on a computer monitor. In a new development, photographic plates composed of solid-state silicon detectors enable the generation of images that are sharper and clearer than traditional X-ray pictures while subjecting the patient to less radiation. Imaging methods like computer assisted tomography (or CT) and magnetic resonance imaging (MRI) can reveal detailed cross-sectional representations of organs deep inside the body. Positron emission tomography (PET) can track metabolic processes in a living organism, even to the extent of revealing the location of activity in the human brain when the subject is asked to perform a particular task. These new medical imaging technologies have largely replaced exploratory surgery that previously would have been used to assess a patient's condition. They have also revolutionized the practice of radiology. "Twenty years ago," says a Connecticut radiologist, "you were dependent on your clinical skills to interpret shadows on an X ray. Now a radiologist's skill is judged by the technology he has learned to utilize."

Computer-mediated pictures of the cross-sections of hearts, brains, and kidneys have become commonplace. But the scientific theory and technological breakthroughs that went into their development are not. The medical imaging techniques of the 1990s are the product of decades-long collaborations among highly trained scientists, mathematicians, engineers, computer scientists, and doctors working at the leading edges of their fields.

## Probing the Body: The Development of Tomography

Although conventional X rays produce clear images of bones and some organs, they cannot give good representations of the heart, located as it is behind the ribs, or the brain beneath the skull. One technology that

## X-RAY BASICS

Images produced by CT, MRI, and ultrasound appear to be similar when viewed on a television monitor, but are, in fact, generated by different methods. These technologies aim electromagnetic waves—X rays, radio signals, or sound waves—at the object to be scanned, while detectors measure the resulting path of the ray or wave and computers manipulate these measurements to create an image. In a conventional X ray, for example, millions of highly energetic rays are aimed at the object; the image is formed as the rays travel through the object and strike a photographic plate. In an ultrasound scan, the source is a transducer that produces sound waves that bounce off the object being scanned. Detectors measure the progress of the waves, and a computer then manipulates the data to create images.

**Signal-to-Noise**
The clarity and accuracy of a medical image depends upon the signal-to-noise ratio of the imaging technology used. This ratio reflects the extent to which rays bounce around randomly before striking film or a detector, thus compromising the quality of the image obtained. In a conventional X ray, for example, some rays make it all the way through the object being examined in a straight line. These rays produce a black imprint on the plate. Others of the rays, known as *attenuated rays*, are stopped when they encounter a substance, such as lead, or bone, that absorbs them. The areas where those rays were absorbed show up as white areas on the plate, which is why bone shows up as a white silhouette. But some portion of the rays ricochet off tissue and bounce off in random directions, leaving gray blurs—noise—on the plate. These are known as random rays, and the proportion of image-producing attenuated rays to random rays is known as the signal-to-noise ratio.

solves this problem is called *tomography*. In this process, organs or bones that are in front of the target are blurred out, and pictures that are "slices," or tomographs, of a particular organ are developed. The image is obtained by moving both the source of the rays and the X-ray plate in opposite directions to each other, while focusing on a particular plane, or point, in the body. It is as if one were taking a picture of a specific tree in a forest—a tree hidden by other trees. As one focuses on the target tree and moves the camera, the trees that are in front seem to disappear.

The basic method of obtaining deeper images of the body was actually developed as early as 1914, when a Polish doctor, Carol Mayer, was able to get an image of the heart by blurring out the shadow of the ribs. While to-mography continued to be used in subsequent decades, this imaging method only reached its full potential with the coupling of X rays and computers, which had the capacity to reconstruct images from huge amounts of mathematical data. In fact, the inspiration for using computers to reconstruct images of the interior of the human body came from the Apollo space exploration missions of the late 1960s. During the mission, photographs of the moon's surface were taken in strips, and computers were later used to assemble these strips into a complete image of a particular area.

## Organs by the Slice: The Computed Tomography Scanner

The modern CT scanner is shaped like a doughnut, with the subject of the scan lying on a platform in the middle, called a gantry. Inside the doughnut is a CT scanner that rotates around the body. While rotating, the scanner emits millions of beams of X rays. These rays pass through a lead screen that absorbs all of the radiation except a fine "fan" of beams, which then pass through the patient to an array of detectors. Although both the scanner and the detectors move in opposite directions to each other, the beams are focused onto particular points of the body. These points are called a plane; the resulting image, called an axial scan, is literally a slice of the body, be it of the brain or the abdomen. To obtain another picture in the series, the patient is moved on the gantry by a few millimeters or centimeters, and another scan is taken. This process continues so that a series of scans is developed.

Computers are used to reconstruct the data collected by the detectors into a medical image. The data is obtained by measuring the intensity of both the entering and exiting X-ray beams. By dividing the second number into the first, the computer measures the extent to which the X-ray beam was absorbed by the tissue at a particular point. The system then reveals an anatomical slice by mathematically reconstructing the data the computer has received. The computer can also manipulate the image, enhancing it, coloring it, or making it larger or smaller. The computer's program then turns the information into pixels that are displayed on a video monitor.

CT produces detailed pictures of bones and cartilage, as well as accurate images of the brain. Although many radiologists consider magnetic resonance imaging the best method of imaging the brain, a CT scan can spot a brain tumor, and also enables surgeons to save lives because they can quickly diagnose such injuries as trauma, fractures, internal bleeding, and swelling. CT scanners have also enabled the development of interventional radiology. Using the detailed images of tissue provided by a CT scan, a radi-

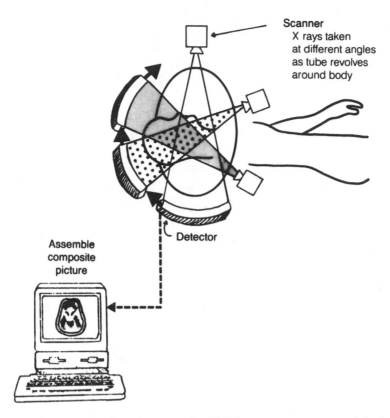

**Figure 7.1** A computerized axial tomography (CAT) scanner rotates around the body, sending millions of beams of X rays through the body to an array of detectors. The scan produces an image that is literally a "slice" of the organ being imaged.

ologist can place needles and catheters precisely into a diseased area of tissue. Injections can be trained through the skin without subjecting the patient to surgery, and, guided by CT, tumors can be located and treated.

CT scanning is also used to precisely align the pins and hole in hip replacement surgery that is performed by a robot, called ROBODOC. In conventional hip replacement surgery, doctors cut a hole in the femur, or thighbone, to insert a prosthesis that is fixed in place with an adhesive. The manual tunneling usually produces a loose fit between the hole and the prosthesis. Combining the CT scanner with the steel nerves of a robot has enabled the development of custom implants that require far less adhesive and are expected to last much longer. Image-guided robotic surgery is also being used in Great Britain to perform prostatectomies on patients suffering from enlarged prostate glands.

The basic principles of CT scanning have been given a new twist in a technology called *spiral scanning*. In a spiral scan, the patient on the gantry moves steadily through the center of the scanner at a set pace. The X-ray source and the detectors continue to rotate around the patient in opposite directions. The effect of the moving gantry is that the X-ray ring follows a spiral path that resembles a spring. How tightly this spring is coiled depends upon the speed at which the gantry moves. This speed sets a variable called the "pitch" of the scan. Because the patient is moving, the spiral CT collects anatomical information quickly and continuously. The technological advance from conventional CT to spiral CT has been compared to the development of motion pictures from still photography.

Spiral scanning is very fast: a single scan can take less than a minute, while a traditional CT scan can take ten or fifteen minutes. There are many advantages to fast scans. Trauma patients, whose lives often depend on fast medical action, can be scanned quickly. Young children or very old patients who have difficulty lying still also benefit. In a conventional scan, if the patient moves, the scan is blurred or has false images, known as artifacts. Speed helps minimize this effect. In the lungs, a conventional scan can miss evidence of small nodules or lesions if the patient breathes between slices. Spiral CT images are collected while a patient holds his or her breath once, so the movements of breathing do not affect the scan.

Fast scanning is also very useful for angiography, the imaging of blood vessels to detect such blood vessel problems as narrowing or blocking of the arteries. Doctors usually use conventional X-ray technology to image arteries. To obtain the image, they inject a chemical, called a contrast medium, into the bloodstream so that the vessels show up in an X ray. Because spiral scanning is faster, a smaller amount of contrast medium is used, which has both cost and medical advantages. As soon as a contrast medium is administered intravenously, the scan begins. CT angiography can detect conditions such as narrowing of important vessels like the carotid or renal arteries. It can also be used to detect tumors in the liver, for example, by scanning the organ while the blood that has been injected with a contrast medium is pulsing through. Different kinds of liver tumors show up at different phases of the scan when a contrasting medium is used. This method of imaging can eliminate the need for a liver biopsy.

Spiral scans are used to produce two types of image. One, which resembles a conventional CT scan, is of a single, discrete slice of anatomy. Using a spiral scanner, this kind of image can be obtained in just a few seconds. The other is a continuous set of data, which is collected while the patient moves through the scan. This kind of scan produces a high volume of data that is then manipulated and reconstructed by computers to produce three-dimensional images of internal organs. The resulting image can be rotated, magnified, or subjected to a close focus in a few seconds. The computer can also add color to images based on varying amounts of X-ray absorption, so that different kinds of tissues are immediately distinguishable. The 3-D visualizations aid in understanding the structures of diseased tissues and their relations to the surrounding normal anatomy. Such images can be used both to aid diagnosis and to plan surgery. In advanced spiral scanning systems, computers can work other wizardry: One system is capable of stripping out both the skull and brain matter from a head scan so that the radiologist can inspect the condition of the blood vessels that surround the outside of the brain. Another computer-mediated system, called 4D CT endoscopy, allows a radiographer to "fly through" the interior of a blood vessel or colon in color almost in real time. The same kinds of computer graphic systems that give moviegoers the sensation of flying through the center funnel of a tornado enable the physician to take a voyage through a patient's vascular or digestive system, without actually inserting a tube into the organ. Parallel computers, which can digest and manipulate enormous amounts of data, process the scanning information to provide this moving endoscopic view.

It is emblematic of the rapid rate of development in medical imaging technologies that even the high speeds of spiral scanning have been surpassed by another method of CT scanning that utilizes electron beams generated by an electron gun to obtain images in milliseconds. One method of electron beam scanning, called *multi-slice mode scanning*, can obtain images from an 8-cm length of the patient's body in 50 milliseconds. The data obtained can be processed by computers to provide a video of the heart, or of a joint, in motion. The electron beam scanner can also be used like a spiral scanner to obtain a three-dimensional image of an internal organ within the space of 100 milliseconds, fast enough to obtain an image of a moving organ, like the heart, while it is "between beats." This kind of ultrafast scanning has proved to be highly effective for obtaining images of the blood vessels around the heart without the need for a more invasive coronary angiogram. The electron beam is particularly suitable for measuring the extent of calcium deposits in blood vessels. These calcium deposits are a symptom of plaque deposits in an artery, and are a good predictor of heart disease even before a patient has experienced any symptoms. However, although electron beam scanning is very fast, the quality of images obtained is not as high as those produced by spiral or helical scanning. "You can only push physics so far," says T. J. Matarazzo, an expert in ultrafast scanning technologies.

## Imaging Meets Particle Physics: Magnetic Resonance Imaging

While the CT scanner was born from the marriage of computers and X rays, the MRI, probably the most effective method of imaging soft tissue deep inside the body, had different origins. It, too, would be inconceivable without computers. But its scientific foundation was in the area of particle physics and grew from the suggestion of the Austrian physicist Wolfgang Pauli, in 1924, that the protons or neutrons inside the nucleus of an atom would, under certain conditions, become magnetic. An American physicist, I. I. Rabi, called this phenomenon nuclear magnetic resonance, or NMR. (In fact, MRI was called NMR initially; the medical procedure was renamed because of the negative connotations attached to the term "nuclear.")

To obtain images of the soft tissue, an MRI machine exploits the magnetic properties of the hydrogen atoms in our body. As a component of water and fat molecules, hydrogen is the most common element in our tissues. When exposed to a magnetic field, the hydrogen atoms line themselves up like compasses. The "tube" that an MRI patient lies in is effectively a gigantic electromagnet. While in this magnetic field, an MRI machine measures the response of the hydrogen atoms in our tissues to pulses of radio signals. In the presence of the radio frequency of the particular atom—called its *resonance frequency*—the protons in the nuclei change their orientation. As the radio signal is switched off,

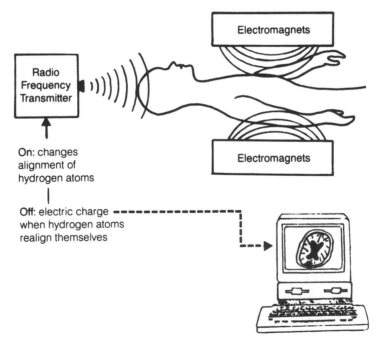

**Figure 7.2**   In magnetic resonance imaging (MRI), detectors measure how hydrogen atoms in the different tissues of the body respond to a radio signal while exposed to a strong magnetic field. The varying responses are used to produce highly detailed images of soft tissue.

the protons produce an alternating signal that can be picked up by a receiver.

Although we think of the water molecule as simply consisting of two atoms of hydrogen and one atom of oxygen always joined together in the same way, the fact is that these molecules are organized differently, depending upon the body tissue they are part of. There are also varying amounts of water in different kinds of tissues. This diversity is reflected in the response of the hydrogen atom to the radio signal of an MRI machine. To create images of various tissues, a computer utilizes data obtained from measuring the differences in the amount of time taken by excited atoms to return to equilibrium from point to point. These measurements reveal subtle differences in adjacent tissues. Thus, brain tissue that is starting to respond to the oxygen deprivation that results from a stroke will appear different from healthy brain tissue. Similarly, a cancerous tumor will appear different from neighboring tissue.

Sophisticated computer algorithms, similar to those developed to recreate pictures from data obtained in a CT scanner, are used to develop images of scanned tissue. As in advanced CT scanning, computer graphics software can manipulate the information obtained from an MRI scan to produce three-dimensional images. However, the different method of obtaining an MRI scan gives it advantages over a CT scan. An MRI scan can, for example, reveal the effects of a stroke on tissue before the damage becomes permanent. A CT scan, however, will reveal only the area of stroke damage when the brain tissue has started to die, as the change in density of the tissue alters the path of the X rays passing through it.

An MRI patient is exposed to 10,000 to 80,000 times the magnetic field we are exposed to in our everyday life. Although there are some concerns about the safety of exposing people to strong magnetic fields, researchers contend the fields are harmless. They argue that life evolved in a magnetic field, albeit the much weaker one of Earth. One precaution that is taken in the strong magnetic fields generated by an MRI machine is to ensure that no metallic objects are present within the magnetic field of the scanner. If they are, they will literally fly into the magnet. Patients with electronic implants, such as pacemakers, also need to exercise caution, because the magnetic field of the machine can distort the electrical functioning of the device. To generate the powerful magnetic fields required of MRI, developers turned to another phenomenon of modern physics: superconductivity (see chapter 8). The magnets in an MRI machine are made of low-temperature superconducting materials cooled to frigid temperatures by liquid helium encased inside insulators. In development are MRI machines that use high-temperature superconducting materials, which can be cooled to their superconducting temperature by less costly liquid nitrogen.

MRI technology is developing on several fronts. Experts are taking advantage of advances in the study of electromagnetic waves and computer science to increase the speed and improve the signal-to-noise ratio to obtain sharper images. Exploiting advances in high-temperature superconductivity, engineers are developing less costly magnets using liquid nitrogen instead of liquid helium to attain the necessary temperatures. Already, advances in the speed of the computer processing of MRI data have enabled the development of MRI-facilitated surgery called *interoperative MRI*. The "operating room" inside an interoperative MRI, rather than being ring-shaped, is in the shape of two U's with a gap of several feet in the middle. Surgeons operate in the gap, usually upon the brain. Guided by virtually real-time images on a monitor, a surgical team, operating with nonmetallic tools inside the MRI machine, can actually see their instruments inside a brain tumor as they work to remove it.

## MRI in Motion: The Development of Functional MRI

While a regular MRI image can provide a sophisticated picture of soft tissue, it cannot show the organ functioning. To accomplish this, researchers have developed a technique called *rapid* or *functional MRI* (known as fMR)—which is capable of taking thirty to one hundred frames per second, a speed that enables MRI to capture biological or metabolic processes in the living brain. A motion picture, by contrast, is projected at the rate of twenty-four frames per second. The method of obtaining fMR images is called *blood oxygen level dependent,* or BOLD, imaging. It was first discovered by Seiji Ogawa, a physicist who published a paper in 1989. He based his research on the assumption that an active brain cell will use more oxygen than one that is at rest. When it is performing a task, the brain cell summons oxygenated blood, which then becomes deoxygenated. This venous blood, lacking oxygen molecules, changes the magnetism of atoms around it when it is placed in a strong magnetic field. This distortion of the magnetic field, in turn, affects the magnetic resonance of hydrogen protons in water molecules nearby, amplifying their signal as much as 100,000 times.

Researchers use fMR to map sites of brain activity. They have been able to pinpoint areas of brain activity as small as 1 mm in diameter. This is equivalent to mapping a football field in six-inch units. The technology is also being used to study thought processes by tracking cerebral events that come and go in milliseconds. FMR generates large quantities of data very quickly, and requires enormous amounts of computing power to analyze it. The technique is not yet in medical practice, but researchers believe that with the use of supercomputers, fMR will become a real-time clinical tool.

## PET: Positron Emission Tomography and Nuclear Medicine

Positron emission tomography (PET) is a different type of technique. It creates images by measuring the paths followed by swiftly decaying radioactive particles as they pass through the body. A key component in the invention of PET was the development of radioactive tracers that are swallowed, inhaled, or injected into the body. The tracers, more accurately called radiopharmaceuticals, are constructed by attaching an isotope (an atom of an element that is unstable because it has extra neutrons) to a biological substance, like glucose, that will move through the body. These radiopharmaceuticals participate in the biochemical and physiological processes of an organism in the same way as the chemicals they replace. They are absorbed at the rate at which the organism needs a particular nutrient, and are cycled through the organism and excreted as the nutrient is metabolized.

The PET system makes use of instruments capable of detecting isotopes from signals that they emit as they move through the body. Computers process the information gathered by these instruments to develop tomographic pictures. The sources of the signals used to develop a PET scan image are particles that make up an individual atom. These particles are called positrons. First detected in 1933, positrons are the mirror image of electrons. While electrons have a negative charge, positrons have a positive charge. Both particles have the same mass. Because isotopes are radioactive, they decay, or lose their radioactivity, over a certain period of time. While an isotope is decaying, it emits positrons, protons and electrons. The positrons and electrons collide with each other, and are annihilated. In the process, two photons or gamma rays are produced. These gamma rays shoot off at 180-degree angles to each other, a phenomenon that is the key to a PET scan.

A ring of electronic detectors, connected to a computing system, surrounds the body, which has absorbed radioisotopes. Whenever two detectors at opposite sides of the ring are hit by photons at the same time, the computer system concludes that they must have been emitted from an isotope decaying inside the body. Then the computer reconstructs a picture of the area where the radioisotope is located.

The PET image itself can be compared to a graph, in which measurements taken are plotted on a two-dimensional map on a computer monitor. Each pixel in the computer monitor reflects such measurements as the volume of the radioisotope in a particular tissue and the rate at which the radiopharmaceutical substance moves through the body. This allows physicians to watch the flow rate of blood or the position of a tracer in the heart or lungs. In fact, radiopharmaceuticals can be used to scan almost every vital organ. They can reveal the way different parts of the brain use energy in a mental process like doing a calculation, pinpoint the location of a can-

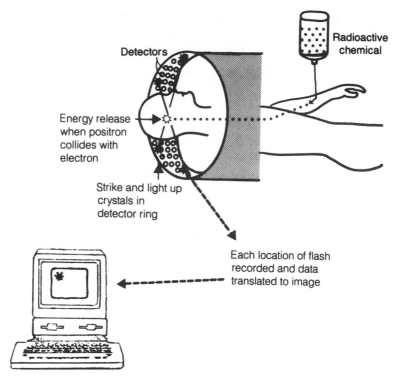

**Figure 7.3** The positron emission tomography scanner can track the way the body metabolizes (uses) a particular nutrient, such as iodine or glucose, that has been modified so that it is radioactive. The scanner pinpoints the location of the nutrient because it can detect the gamma rays that are emitted in the process of radioactive decay.

cerous tumor that has absorbed the isotope in differing amounts from the surrounding tissue, or track the progress of metastatic cancer cells.

The *T* in PET stands for *tomography*, in that PET makes images of planes through the body. But, although a final PET image may resemble other scans, unlike either MRI or CT, the aim is not to produce images of anatomical structures of the particular slice of the body being observed. Instead, PET tracks metabolic functions by highlighting "hot" areas where the radioactive isotope has been absorbed. The computer provides the anatomical images, incorporating the information from the PET scan into the computer image. Recent developments in medical imaging use computers to combine the data from CT and PET scans to produce a representation that combines both anatomical and functional information.

The range and usefulness of PET is expanding rapidly as researchers develop new radiopharmaceuticals that can target specific tissues. Using isotopes that home in on specific neuroreceptors in brain tissue, for example, researchers have increased their knowledge of Alzheimer's disease, depression, anxiety, and schizophrenia. PET scanners also aid in the treatment of malignancies when radioisotopes are used to kill the cells in a cancerous tumor (and, in fact, radioisotopes that target tumors are in development). A PET scan can also provide information about the chemistry of the tissue around a tumor, and can be used to detect malignancies before any tissue changes are visible.

One disadvantage of PET scanners is that they are expensive and require specialized equipment to generate the isotopes. For that reason, *single photon emission computed tomography,* or SPECT, is more widely used to study brain function, even though it is not as accurate or specific. In SPECT, a nuclear agent is injected or inhaled and travels to the brain. As the isotope decays, it emits single photons that are recorded by a specialized gamma camera. The procedure is used to diagnose and study Alzheimer's disease, and to spot brain tumors that an MRI scan might miss. Because the SPECT scan is based on a single event, it is considered less accurate than PET, in which the coincident emission of two photons traveling in opposite directions is measured.

## Ultrasound

Ultrasound uses sound waves, rather than radiation or magnetism, to extract images from the interior of the body. An ultrasound machine develops an image of the inside of the body by manipulating and analyzing sound waves as they bounce off surfaces and echo back to the sender. Images can be created because sound waves travel through different substances of different density at varying speeds. Their speed is also altered by variations in temperature. Ultrasound images are not as sharp as those obtained by other means, but they are popular because they are inexpensive, and can be obtained quickly. Also, they do not use radiation from the high-frequency end of the electromagnetic spectrum, which can be harmful.

As in other imaging technologies, the combination of computers with ultrasound has dramatically improved its range and quality. Using devices called *computerized digital-scan converters,* ultrasound machines can now register even the weaker echoes that come from the spaces between internal organs and record these echoes as varying shades of gray. Because of these advances, real-time imaging has become routine—enabling many an expectant parent to observe the movement and heartbeat of a fetus months before it has emerged into the world. New ultrasound machines are even starting to rival CT and MRI in their ability to produce high-resolution images of the interior of the body. Ultrasound scans that can be viewed on color monitors are in development, as are modified machines that can probe for small tumors by measuring the stiffness of tissue. Portable ultrasound machines that are capable of transmitting images from remote locations, like the battlefield or an accident site, are being tested.

Medical imaging devices use high-frequency ultrasound to obtain images. Low-frequency ultrasound, meanwhile, is being exploited in other areas. In an eye surgery procedure called facoemulsification, low-frequency ultrasound is used to break up cataracts and emulsify them so that they can be removed by vacuum suctioning. Ultrasound is also used to break up kidney stones, a procedure called shock wave lithotripsy. The precision of a device called a *cavitron ultrasonic surgical aspirator* (CUSA) is exploited to remove brain and spinal cord tumors while inflicting minimal damage on surrounding tissue.

## Images of the Future

With the rapid advance of imaging technologies, applications that may have appeared highly improbable a decade ago—such as performing surgery inside an open MRI machine, or transmitting complex medical images over telecommunications lines—are already a reality. Because imaging machines are very costly—an MRI machine, for example, can cost more than $1 million—and the market for new and expensive equipment is limited, changes in imaging hardware are not rapid. Advances in imaging technology are likely to occur more quickly in such areas as computer software and data communications. One possible future development in MRI technology will be the combination of MRI with sound waves to simulate touch as well as sight. This approach allows physicians to "feel" hard cancer tissue on a computerized image with virtual reality tools—a technique dubbed *magnetic resonance elastography*. This technique could be used for remote diagnosis (see page 181), and also in situations where the tumor is located deep inside the body. Another developing technology, *magneto-encephalography* (MEG), is expected to enable major advances in brain imaging because of its sensitivity. MEG uses an array of superconducting sensors (SQUIDs) to measure brain waves that are so subtle they would not be registered by other scanning methods (see chapter 8).

Researchers are also working on developing new kinds of X rays with improved signal-to-noise ratios. This is especially important for the diagnosis of cancers, which are often very subtly different from normal tissue, and thus difficult to diagnose. Standard X-ray tubes bombard patients with a wide spectrum of X rays. These range from "soft" rays that barely penetrate skin or tissue and do not produce an image on the X-ray film, to "harder" rays that hit tissue and bounce around before registering on the X-ray film. These rays create a fog that can obscure tumors. Between these two extremes, there is a frequency that passes through tissue and registers on X-ray film with little or no scattering. Because laser light consists of a coherent and monochromatic beam of electromagnetic radiation, researchers propose using a *free electron laser* (FEL) to generate the desired frequency. In studies, researchers using the free electron laser facility at Vanderbilt University in Memphis, Tennessee, have found that cancers absorb about 11 percent more X rays than normal tissues. This different rate of absorption can be used to generate a very detailed image. Once the tumor is detected, doctors could use the image to determine whether the growth is benign or malignant. Also, because the monochromatic beam uses only a fraction of the number of X rays produced by traditional radiology, it exposes patients to considerably less radiation, reducing the risks of regular mammograms and other types of cancer screenings.

In addition to techniques that let physicians see more clearly inside the body, computers are also helping radiologists interpret the images they see. Software incorporating artificial intelligence and machine-vision algorithms can scan mammograms and chest X rays for telltale signs of cancer. These kinds of computer-aided diagnostic tools offer an automated "second opinion" that has proven particularly useful because computers analyze images differently from the ways people do. The patterns a computer programmed with machine-vision algorithms can easily identify are not the same as those noted by the human visual system. Thus, computers can complement the radiologist's eye, a development that may be of particular benefit in mammography, where the signs of a tumor are often quite subtle and can be missed.

## High-Tech Flashlights: Fiber Optics Explores the Human Body

Advances in medical imaging can replace the use of exploratory surgery to assess, for example, the location and condition of a cancerous tumor. Using the same kind of ultrapure glass fiber that has dramatically increased the capacity of the international telecommunications network, a doctor can examine and even perform medical procedures in the deepest and most convoluted recesses of the body. Rather than cutting through layers of tissue to reach the area, the surgeon inserts the tube through an orifice, or through a tiny incision in the body. Although fiber-optic telecommunications cable can be thousands of miles long, its medical counterpart is a flexible fiber-optic tube equipped with a light, and sometimes a tiny camera, that ranges in length from about one to five feet. The use of a fiber-optic system permits the doctor to see the area being examined on a television monitor or using a special viewing aperture. The doctor can also pass different types of instruments through a small channel in the tube, or the tube can be equipped with a set of miniature tools, such as a small clipper to remove tissue samples, or an electric probe to destroy abnormal tissue. The endoscope can even be used to transmit laser light to destroy malignant tissue that has been infiltrated with a light-activated drug (see chapter 6).

*Endoscopy* is the general term for examining the internal organs of the body using a fiberoptic viewing tube. The term is also used to refer to the procedure of using an endoscope to diagnose and treat digestive system disorders. This instrument consists of multiple hairlike glass rods bundled together into a tube that is less than half an inch in diameter. The fiberglass tube transmitting the light is specially fabricated to keep light within its tube even as it bends and twists. Thus, the intense light transmitted by the system enables the endoscopist to see around corners as well as forward and backward. There are many medical variations on endoscopy. *Hysteroscopy*, for example, uses a fiber-optic tube to see directly inside the uterus. *Bronchoscopy* is the use of a fiber-optic visual system to study the airways of the lungs for a tumor or other abnormality.

In *laparoscopy*, a thin viewing tube is inserted into the abdominal cavity through a small cut made at the bottom of the navel. Carbon dioxide is used to inflate the abdomen so that organs in the entire abdomen and pelvis can be seen clearly. Laparoscopy is used for both diagnosis and treatment. It is often used to determine the cause of such gynecological problems as pelvic pain or sterility. It can also be used with other instruments that are inserted through other small incisions in the abdomen to perform procedures like biopsies and sterilization. And, as the field advances, the range of abdominal procedures performed using laparoscopy is steadily increasing. Valuable diagnostic information can be obtained by examining a biopsy specimen of the liver or of abdominal lesions. Surgeons also can perform a variety of procedures with this method, such as removing the gallbladder.

In orthopedic surgery, the same technique is called *arthroscopy*, and it simplifies the treatment of many disorders that previously required a large surgical incision and a lengthy period of rehabilitation. An arthroscope gives an orthopedist a clear view of the cartilage inside a damaged knee, shoulder, or other joint by training light into the area through the optical fiber. The lenses in the scope form an image that the camera sends to a television monitor. To expand the area around the joint and provide space for the orthopedist to work, a *cannula*, which is a type of drainage tube, is used to transport a salty solution into the area of the joint. Miniature tools repair the damage to the joint by shaving away torn meniscus cartilage or mending torn ligaments.

Using endoscopy, surgeons can conduct procedures that previously were impossible because their scale was too small. Also, because these techniques are less invasive, patients recover more quickly from surgery; some procedures that previously required lengthy hospital stays are now outpatient procedures. Because of its facility for probing deep recesses of the human body and transmitting the accumulated information into computer monitors via small cameras, endoscopy is expected to become an important element of the remote medical practice that has been dubbed *telemedicine*. It can also be an important adjunct to robotically assisted surgery. In fact, heart surgery using robotically assisted endoscopic tools was recently performed for the first time.

## Medicine on the Information Superhighway: Transforming Health Care in the Information Age

Advances in digital technologies and communications have led to the development of another growing area of high-tech medicine: the same data networks that can transmit credit card information can send medical data from one end of the globe to the other. The information transmitted over global networks is potentially any kind of data that can be turned into the ones and zeros of digital information streams, including medical images and blood chemistry measurements. In future decades, virtual-reality information may be carried over these networks. For example, using a "data glove," a doctor will be able to "touch" the mass in a patient's stomach from thousands of miles away. The general term for using data networks to deliver medical care is telemedicine. It involves the collaboration of doctors, computers, robots, imaging devices, and electronic communications. In the United States, it is an evolving field with many of its pieces already in place. These pieces include "smart cards," which can record a patient's medical history on a magnetic strip, and hand-held computers equipped with radio communications devices. Specialized equipment, like high-resolution cameras called Medicams, that can transmit thirty frames a second, have also been developed. Other elements include medical instruments, like a blood pressure monitor, that can collect patient data and can be linked into telecommunications networks, as well as digital libraries and databases that are becoming increasingly easy to use.

Although not yet used extensively, telemedicine is already a boon for patients who do not have ready access to health care because of their location or their circumstances. International telemedicine can supplement the health care available in other countries, and there have already been reports of its use to provide a long-distance diagnosis—using X-ray images transmitted by satellite or over the Internet—for patients in third-world countries. In one pilot program, doctors in Armenia treating the victims of a massive earthquake in that country used satellite links to confer with doctors in the United States about patients' diagnoses and treatment protocols. Although such satellite communications as those used in this project would be prohibitively expensive for regular use, the practice of long-distance medicine is becoming increasingly more practical as the cost of satellite communications drops. Though at present the relatively slow speeds and congestion of the Internet make it an unsuitable medium for "live" telemedicine hookups, this is expected to change as international fiberoptic or satellite communications links are developed. The tools and technologies of telemedicine are also being used to simulate conditions for medical students and practitioners to examine virtual patients, to access online databases, to carry out remote consultations, and even to practice complex procedures using virtual-reality tools.

---

**TAKE TWO ASPIRIN . . . : A TELEMEDICINE GLOSSARY**

In telemedicine, the prefix *tele-* usually denotes the various kinds of medicine practiced. A radiologist uses teleradiology, for example, to view transmitted medical images on his or her computer monitor. *Telepresence surgery* is surgery performed by a distant physician with the help of a robotic arm, or using a tiny robot to conduct an internal physical exam. (These procedures use the same kinds of fiberoptic tools as those used in endoscopy.) *Telepathology* is the distant diagnosis of an infectious disease; it also describes efforts to use remote sensing to improve health conditions around the world. *Telehealth* is defined as the provision of remotely located health information or services. *Telementoring* is providing expert medical guidance at a distance. The kind of mentoring that is envisioned is of a remote specialist assisting in the diagnosis and treatment of a patient by a local internist using such tools as live video and other monitors of body functions.

---

Various telemedicine projects are already in operation in the United States as well as internationally. A comprehensive telemedicine system is in development in the state of Georgia in the United States, where many remote communities have little access to specialized medical care. The Georgia Statewide Telemedicine Program (GSTP) links hospitals, health departments, community and public health centers, and state prisons in towns and cities across the state. Ultimately the GSTP will link fifty-four telemedicine sites. This project uses high-capacity telephone lines to transmit fully interactive two-way audio, video, and diagnostic information. The sites are equipped with high-resolution cameras called Medicams, which can zoom in and focus on a patient to allow detailed examinations. Microcameras can be connected to instruments that examine ears and eyes. A patient's heart and lung sounds can be transmitted to the consulting provider with the use of an electronic stethoscope. Additional external video cables allow live video from other equipment, such as an ultrasound machine or other image-generating diagnostic devices, to be transmitted over the system. A data management system keeps an electronic telemedical record of the telemedicine consultation. To date, more than two thousand telemedicine consultations have been conducted.

In a test of remote care, the Georgia team in 1994 installed home health care systems in the homes of twenty-five people with chronic conditions who were constantly being readmitted to the hospital. The residential units were linked to hospital-based units using a two-way coaxial cable connection similar to that used in cable TV. The system combined interactive audio and video conferencing between patients and caregivers with residence-based monitoring equipment to measure patients' pulse rate, blood pressure, and temperature, as well as blood chemistry. The equipment included electronic stethoscopes and even ultrasound equip-

---

## MONITORING IN MOTION

The same communications technologies that deliver medicine at a distance are being used to develop a new generation of high-technology medical transports called Life Support for Trauma and Transfer. These kinds of vehicles are intended to help save the lives of seriously ill or injured people. It has been found that the highest fatality rate in an accident or battlefield conditions occurs in the first thirty minutes after trauma. One such vehicle is equipped with an integrated computer system that constantly transmits data on the patient's status from the carrier to the hospital, using monitors that measure such life functions as pulse rate and blood pressure. The ambulance is also equipped with a miniature respirator, an oxygen supply, temperature controls, environmental regulators, and filters to protect against biological or chemical contamination. In the future, an ambulance arriving at an accident scene might also have a device called a *portable direct digital imager*, a small, filmless X-ray machine that can transmit images from the field to the specialist.

184 • Who Gives a Gigabyte?

ment for assessing fetal heart rates and uterine activity. The goal was to improve the management of chronic medical conditions and reduce the need for patients to make arduous and costly visits to the hospital. Because of the project's success, it has been extended and will be commercialized.

Other ambitious applications of telemedicine are being developed. In 1988 a team of telecommunications experts, mountain climbers, doctors, and technologists made their first attempt to implement a telemedicine project from the top of the world. This test of communications technologies and newly developed medical monitoring devices was conducted on Mount Everest, one of the most extreme and remote environments on Earth. Expedition leaders contended that if the cutting-edge medical technologies worked on Everest, then the possibility exists to deliver state-of-the-art medical care anywhere.

The plan for the Everest Extreme Expedition was to monitor the physiology of the climbers from base camp as the four-man team made its ascent to the summit. The four climbers wore tiny sensors developed at the Massachusetts Institute of Technology Media Lab. The sensors included a blood oxygen sensor, worn on the center of the climber's forehead; a heart rate sensor, worn on the chest; thermistors, which measure skin temperature under each arm; and a body core temperature monitor, which was swallowed like a pill, to transmit body temperature readings. The sensors were equipped with tiny transmitters to relay the physiological data back to base camp.

At the base camp, the data was to be processed by computers designed to withstand high altitudes. From the base camp clinic, video and data were transmitted via an INMARSAT satellite in geostationary orbit over the Indian Ocean to a land station in Malaysia. From there the calls were routed over a global telecommunications network to California, and then to an ISDN network. Using the ISDN network (see chapter 3), calls were linked to a video bridging service that enabled multiple sites to monitor and participate in video calls.

The outcome of this ambitious telemedicine project illustrates some of the problems that one confronts in implementing any advanced technology. Data from the climbers, who did successfully reach the summit, were not satisfactorily transmitted from the climbers to the base camp. Moreover, some of the climbers found their biosensors cumbersome, and removed them. Communications between the base camp clinic and the United States proved more successful, however, and several telemedicine conferences were conducted. Images from a portable three-dimensional ultrasound unit that was also being tested were successfully transmitted from the base camp to telemonitors in the United States.

In the more distant future is *telesurgery*: the use of remotely operated mechanical arms that cut, suction, and sew. Proponents of telesurgery consider this development particularly promising, because the small machines that can explore, photograph, and transmit photos of the twists and tight spaces in a human body, such as the gastrointestinal tract, are already in use (see page 179). They foresee that these devices could in the future function as peripherals of a telemedicine system just as small cameras now enable the otoscope, microscope, endoscope, and other instruments to transmit images from deep inside the body to a television monitor.

A recent experiment conducted by the Yale University School of Medicine and the National Aeronautics and Space Administration demonstrated the feasibility of remote surgery. Using laptop computers and phone lines, consulting surgeons at Yale, in New Haven, Connecticut, were linked to a mobile surgery program in Cuenca, a town high in the Andes Mountains, and to an isolated hospital ten hours away from Cuenca, in Sucua, Ecuador. The Ecuadorian doctors in Cuenca first collaborated with primary-care physicians in the isolated hospital over the phone. They were able to plan their operations, including state-of-the-art laparoscopic surgery. The team then traveled to Sucua, where the doctors and patients confirmed their plans in the mobile surgical unit in the jungle. A laparoscopic procedure was performed with real-time consultation and monitoring from the Yale campus. Surgeons at Yale guided the surgical dissection and confirmed all critical elements of the anatomy and operative techniques.

The exciting potential of telemedicine has already spawned several government reports and groups that share information on the Internet. Because of its utility for treating soldiers in the battlefield, or who have been posted to remote regions, telemedicine research has received considerable funding from the military, particularly the Defense Advanced Research Projects Agency (DARPA). But the full implementation of telemedicine faces a number of technical obstacles. In recent studies, which compared remote with face-to-face diagnosis, it was found that doctors needed to be far more familiar with the technologies to make reliable diagnoses. In addition, because of its reliance on video and medical images, telemedicine will require high-bandwidth communications networks that are not yet in place. However, the continuing advance of communications technologies will probably resolve many of the technical challenges. A more important practical hurdle—one that is faced by many developing technologies—is whether it will be possible to find financing to develop the tools of telemedicine. Legal and societal issues also need to be resolved. These include uncertainty about exposure to malpractice suits, a lack of guidelines on how to use the hardware and software, and questions about who pays for a medical service delivered from afar. In addition, as in other enterprises that make use of telecommunications, there are concerns about privacy and confidentiality. If these issues can be resolved, the implementation of telemedicine will probably change the way many people make an appointment with the doctor, and will deliver considerable health benefits to people living in remote areas or to countries whose own medical infrastructure is lacking.

## Spare Parts and Repair Kits: New Developments in Bionics

Engineers and materials scientists marvel at the devices and materials that nature produces. The complex machine that is the human body, for example, is designed from strong and flexible materials that renew themselves constantly, are exquisitely engineered to bear loads and stresses, and function for decades at a time. But there are times when the body is unable to repair the effects of an inborn defect in an organ or the ravages of disease, overuse, aging, or accidents. One solution is to replace a diseased or defective organ with a transplant from a healthy donor, but this avenue is restricted by chronic shortages of donor organs. Every year, at least 100,000 Americans die waiting for a replacement heart, liver, kidney, or other organ. Moreover, there are frequent problems of compatibility and rejection, as well as concerns about transmitting infections.

As an alternative to organ transplants, scientists and engineers have turned their toolmaking skills to correcting the medical problems the body cannot repair. Borrowing from materials science, microelectronics, and microbiology, researchers have begun to develop a growing selection of artificial body parts, a wide range of which are already available. Hundreds of thousands of people around the world receive artificial hips and knees, heart valves, tooth implants, spine supports, eye lenses, and other replacements. Other implants help disabled organs or other body components to function. These include cardiac pacemakers and defibrillators to maintain regular heartbeats and rhythms, and internal braces and splints to strengthen weakened or shattered bones. Cochlear ear implants, which replace the neural mechanism within the inner ear, can filter out noise, compress signals, and even select the best frequencies to send to the brain to make it easier to understand a sound. While they do not enable normal hearing, these electronic devices have become so sophisticated that some deaf or severely hearing-impaired people can, with training and practice, process speech well enough to talk on the telephone.

With advances in computer software and materials, artificial joints and limbs can be exactly fitted to meet a particular patient's needs. To build an artificial joint, for example, a bioengineer can select a particular implant design from a database, and then fit the design using a computer monitor and special software that incorporates various kinds of relevant data from the patient. Technological advances in materials and micro-electronics have also created a new generation of artificial limbs for amputees. For example, a microprocessor-controlled prosthetic knee joint adjusts to changes in speed and walking rhythm, letting the wearer walk with a more natural gait. Microprocessors in other prosthetic devices are linked with electrodes to muscles in a thigh or shoulder that control the devices' movements. At Rutgers University in New Brunswick, New Jersey, researchers have developed a device that can sense tendon movement in an amputated arm. A signal generated by an individual tendon can move a single finger on a robotic hand.

Many more new devices are in various stages of development around the world. One area of intensive research is in ways to enhance or replace the human visual system. One device in development, for example, is designed to enhance images for people with limited vision. Inside a pair of goggles are two wide-angle cameras and a zoom lens that feed images to a battery-operated waist pack. After magnifying images and adding contrast, this low-vision enhancement system projects images onto screens mounted onto the goggles. Another, more ambitious plan that is probably decades from being practically implemented contemplates a system in which a tiny computer chip is implanted into the back of the eye. This chip would receive visual data from a laser mounted on a pair of goggles, and relay impulses through the optic nerve to the brain. Other researchers are experimenting with methods of mapping camera signals straight onto the visual cortex of the brain. This will, in theory, enable the patient to "see."

Other exciting research is focused on combining natural cells and tissues with artificial materials to make hybrid artificial organs. An example is the development of an artificial liver that consists of liver cells that are grown in tissue culture and placed into implants connected to the blood supply of the recipient. In development is an artificial pancreas to treat diabetes. Some investigators are testing the possibility of implanting insulin reservoirs into the abdomen that are equipped with sensors that detect blood sugar levels and automatically dispense the appropriate amount of the hormone. At the Georgia Institute of Technology, researchers are developing a specialized membrane that can encapsulate cells that produce insulin. The membrane is structured with pores small enough to permit the glucose to pass into the cells, and allow the insulin to go out. But the pores exclude antibodies and lymphocytes, thus forestalling immune system rejection of the artificial tissue. Another area of great potential is the use of electrode implants in the body to activate functions impaired by damaged nerves or to stimulate parts of the brain and central nervous system to get a desired effect, such as preventing seizures.

## Materials Science Goes Back to Nature: Tissue Engineering Comes of Age

In their quest for body parts that are not hampered by problems of scarcity, infection, and rejection, researchers are turning to a promising new field. Developed only in the past two decades, tissue engineering melds materials science with cell biology to harness and guide the body's own capacity to heal and regenerate tissue even in cell types, like cartilage, that are not normally regenerated. Its proponents predict that tissue engineering will revolutionize the practice of medicine within the next decade. In the future, tissue-engineering technology may enable surgeons to repair damaged spinal cord nerves, or replace an arthritic knee with one made of human bone and cartilage.

Tissue engineering is a multidisciplinary field that has come into its own as a result of converging advances in several sciences. One of these is the growing understanding of the immune response and cell biology; another is the discovery and development of new materials. In the early development of medical implants, researchers and clinicians used biologically inert metals and polymers to avoid the body's tendency to reject foreign natural materials. But this class of implants frequently becomes infected, and can incur immune responses like inflammation and rejection. Then, in the 1980s, researchers began to utilize a class of materials that interact with cells in a tissue in a beneficial way and can be harnessed to help the tissues in an organ heal and regenerate themselves. These biodegradable materials, including collagen and a protein known as chondroitin, can be formed into matrices that interact with cells and provide a kind of scaffolding that encourages the growth of new tissue. Researchers have developed methods of using these scaffolds to regenerate skin, bone and cartilage and they are already in clinical use. Among these products is an artificial skin that is used as a skin substitute for patients suffering from severe burns. The product consists of a biodegradeable material that forms a template that new skin cells can anchor themselves onto, and a protective silicone cover. The microscopic structure of the template has been optimized to encourage the growth of normal skin cells rather than the scar tissue that would otherwise form.

Researchers are experimenting with using similar methods in the laboratory to grow other organs, such as the bladder. In one method, the researcher takes a biopsy from an animal of the type of tissue selected to grow. Next, the researcher extracts the cells from the tissue, using enzymes to digest the proteins, chemicals, and molecules that form the matrix of the tissue. The living cells are grown in a petri dish surrounded by a mixture rich in nutrients. Meanwhile, the researcher makes a supportive framework in the shape of the needed part, using extremely fine, porous polymer mesh. Cells from the petri dish are seeded onto the scaffolding, where they continue to multiply and grow into the desired shape.

Because the structure of an organ is often quite intricate, researchers have developed various methods of creating the basic structure for the cells to grow on. One method uses a three-dimensional printer that was invented to fabricate engine parts. The instrument has a printing head that deposits material on a surface, like an ink-jet printer. But instead of depositing ink on a page, this device deposits successive layers of a material to create a three-dimensional structure.

Other medical researchers have grown replacement organs for sheep, rats, and rabbits, using the animals' own cells and lab molds to help the tissue take shape. This procedure has reportedly been successfully used to grow replacement organs for newborn animals while they were still in the womb. Researchers hope this technique can be used in the future to correct common birth defects as well as to grow new organ tissue in older people.

To perform this procedure, doctors operated on the animal fetus two-thirds of the way through pregnancy. Using laparoscopic methods, they lowered a surgical camera and long, narrow instruments into the womb. Guided by a large video monitor, they removed a small sample of the defective organ. In the laboratory, lab technicians separated different types of cells and placed them in dishes of a clear solution rich in proteins and nutrients. Next, the researchers built the organ by draping the tissue over biodegradable scaffolding. A sheep's bladder, for example, was fashioned by layering epithelial cells, which line our bodily organs, on the inside of a cup-shaped structure and muscle cells on the outside. The researchers found that the cells oriented themselves to each other and grew until they filled out the scaffold. Within six weeks of surgery, the new bladder was transplanted in the newborn sheep, where it functioned well.

Despite these developments, it will probably be decades before tissue laboratories are able to fulfill a surgeon's dream of ordering from a catalog of living prostheses for every organ system in the body. In order to attain that goal, there are several challenges that need to be addressed. One of the main problems is ensuring that these organs are not rejected by the body's immune system. Another is the development of new materials that can be completely degraded and absorbed by the body. Yet other research focuses on the molecular architecture of the matrices—or scaffold—of particular tissues, and the task of reproducing them and seeding them with a blood supply. More investigations are also needed into the chemistry of the matrix that supports cells in a particular type of tissue. (This matrix, known as the *extracellular matrix*, functions as the cement that holds the bricks of our cells together.) Polymers, which are long chains of molecules in the extracellular matrix, have chemical sequences and patterns that provide signals believed to be crucial to normal cell function. Researchers need to gain a greater understanding of this interaction so that they can provide the best conditions to encourage the cells to develop into viable tissues.

# 8

# Material Improvements: Better Living through Advanced Chemistry

In our everyday lives, we are surrounded by thousands of materials that did not exist at the turn of the century. Most people are aware of this: words like *plastic* and *silicon* have acquired connotations in modern discourse that go far beyond their material properties. We are less aware, however, of the roles that materials play in the development and improvement of modern devices and machines, of the extent to which material improvements are a precursor—or an enabling technology—for other advances. Take, for example, machine engines, which operate with considerably greater efficiency at higher temperatures. At the turn of the century, cast-iron engines operated at between 400 and 600°F—the temperature of pressurized steam. By contrast, a jet turbine blade fabricated by growing crystals of combinations of metals, called superalloys, can run at above 2,000°F.

To a large extent, the digital and communications revolution was also a materials revolution. We would not have been able to "wire the world" without the development of ultrapure glass fibers to carry the light signals of billions of bits of information without scattering or loss. The evolution of microcomputer circuits would have been impossible without the discovery that adding a few atoms of elements like boron or phosphorous to silicon could transform this cheap and ubiquitous material into a semiconductor. In the future we will be looking to materials scientists to address such global problems as energy supply, housing, transportation, and communications, with light and strong materials that can be manufactured and can function without exacting a heavy price on the environment.

Traditionally, the work of manipulating materials was the domain of craftsmen and builders whose knowledge was based on long apprenticeships and experience. Their ability was a combination of age-old lore and day-to-day experience, probably informed by some commonsense knowledge of physics and chemistry. Today's materials scientist, however, is likely to have expertise in chemistry and other formerly separate fields such as mathematics, electronics, mechanics, physics, and even biology. In fact, materials science encompasses an enormous range of disciplines from metallurgy to biomimetics (the study of how materials like bone, a tree trunk, or an oyster shell are constructed in nature). The task of developing semiconductors that can pack millions of microscopic circuits into a square centimeter of silicon or gallium arsenide is considered the domain of the materials scientist rather than that of the computer engineer. In this field, advanced knowledge of chemistry, crystal structures, and quantum mechanics is particularly vital for the development of new semiconductor materials for electronic components whose dimensions are approaching molecular levels. Moreover, in a circular relationship that characterizes much of materials science, the semiconductor industry has benefited from advances in methods of designing and producing advanced materials, and has also become the source of methods of developing even more sophisticated advanced materials like microelectromechanical machines (see page 214).

## GUIDING PRINCIPLES:
## THE PERIODIC TABLE OF THE ELEMENTS

To the layman, the periodic table of the elements resembles a mysterious checkerboard filled with numbers and chemical symbols: to a materials scientist, this table is a guide to the materials world that is replete with possibilities. It is actually an organized catalog of all the known elements, characterized by atomic weight, or the average mass of the element, and chemical properties. Currently listing 112 elements by atomic weight (in rows) and chemical properties (in columns), the periodic table serves as an important pathfinder to chemists and materials scientists in their efforts to improve traditional materials and devise new ones, because it groups the elements into "families" that share the same properties. For example, in one column are highly reactive elements called alkali metals, including lithium, sodium, and potassium. Each of these elements forms compounds, called salts, by combining in a one-to-one ratio with any of the halides grouped in another column. These include flourine, chlorine, and bromine. (The most common compound that results from these two columns is sodium chloride, or table salt.) Similarly, materials scientists use the table as a guide in their search for new semi- and superconducting materials (see page 206).

The table was developed in the late nineteenth century by a Russian chemist, Dimitri Mendeleev, who discovered that the known elements could be arranged by order of increasing atomic number (that is, the number of protons in the nucleus of the atom) and that their chemical and physical properties recurred periodically. Mendeleev's research revealed an apparent natural order to the elements that has been borne out by the fact that holes Mendeleev left for elements that he predicted would be discovered have since been filled in. The quantum theory of the structure of atoms explains this observed periodicity: The electrons in an atom are organized in a series of orbits, each of which contains a fixed number of electrons. The number of electrons in the outer orbit determines the chemical properties of an element. Even if they differ in weight and size, atoms that have the same number of electrons in their outer shell have similar properties.

## From Church Bells to High Explosives: Materials through the Ages

Although our modern world includes tens of thousands of new substances that did not exist at the turn of the century, many of the materials that play an important part in our lives have been in widespread use for hundreds and even thousands of years. Metals have been mined and worked for millennia. Modern methods of making metals useful, by hardening or aging, tempering, and beating or molding them into a desired shape, have their forebears in centuries-old practices. Through the ages, material improvements came about as the result of trial and error and the application of expertise from one area to another. For example, although the Chinese are credited with inventing gunpowder several centuries before it was known in the West, these explosives had limited military value because there were no safe and effective gun barrels. The art of casting cannons, developed in the West during the fourteenth century, was an offshoot of the technology for casting church bells. Only then did gunpowder become an effective instrument of war.

Throughout history, in fact, the needs of the military have driven the development and improvement of materials: many ancient wars were won or lost depending on which side had developed harder and stronger materials with which to construct swords and crossbows. In the modern age, huge military budgets have helped underwrite the costly development of advanced materials—especially in aerospace, where every extra pound detracts from the performance of a jet or rocket and adds to the cost of the fuel. Advanced materials also owe their provenance to the materials demands of space exploration, which call for spacecraft manufactured from light and strong materials that can tolerate such enormous stresses as those created by changes in gravity, extremes of temperature, and rocket thrusts.

Whereas a traditional craftsman would have been directed by the outward appearance of a material and its performance in the field, a modern materials scientist is guided by sophisticated tools and methods that probe the nature of materials at the molecular—and even the atomic—level. Surprisingly, however, even though the current generation of materials scientists has extensive scientific knowledge and precise tools at its disposal, many of today's sophisticated materials, like superalloys, advanced composites, and high-temperature superconductors, actually owe their development as much to traditional approaches, including trial and error, as to advanced science. In the future, however, observers predict the advent of a new generation of materials that will be designed and constructed to order, atom by atom.

## ATOMIC-SCALE CONSTRUCTION: MOLECULAR BEAM EPITAXY

Increasingly, advanced materials science involves the manipulation of atoms and molecules of materials. Because of the requirement of etching millions of microscopic computer circuits on a silicon chip with an area of several square centimeters, the semiconductor industry has become expert in developing methods of constructing materials layer by atomic layer. The most commonly used method of "building" a microchip is *chemical vapor deposition*, in which the constituents of a microchip are vaporized in a vacuum chamber and then laid down, layer by layer, on a substrate (or base). An advanced variation of this technique, called *molecular beam epitaxy* (or MBE), "builds" semiconductor crystals using a method that has been described as "atomic spray painting." The typical MBE machine consists of a stainless-steel vacuum chamber in which a crystal substrate rests on a platform that has been set at a specific temperature. Crucibles containing various chemicals or minerals, such as gallium, arsenic, aluminum, and indium phosphide, are separated by shutters from the main chamber. When the crucibles are heated, these substances emit atoms, which then turn into gases. Computers open and close the shutters in nanoseconds to control the movement of the gases into the vacuum chamber. In the chamber, the substances form a layer on the crystal that can be just one layer of atoms thick. By manipulating this spacing between new combinations of elements, materials scientists hope to develop faster electrical components and crystals with exotic properties.

MBE has already made its way into our everyday lives: this deposition method is used to manufacture more than 70 percent of the semiconductor crystals that generate the laser light used to read the surface of a compact disk. Lasers constructed using MBE are made of compounds such as aluminum gallium arsenide and gallium arsenide layered upon one another, with each layer just a few atoms thick. The sandwich structure of these crystals causes a quantum mechanical effect, the input of electrons causes electrons moving within the crystal to meet with positively charged vacancies in the crystal. Laser light is generated between the crystals at the point where the electrons and positively charged vacancies—known as "holes"—meet (see chapter 4).

## Shake 'N' Bake: Developing New Materials

Many researchers believe that humanity has barely tapped the diversity of materials that can be developed from the elements that we have at our disposal. And they dream of constructing new materials with properties that are tailored to meet exact specifications. In some areas, such as polymer science (see page 218), chemists have already developed such an extensive database of knowledge about these giant molecules, they are virtually capable of designing a synthetic polymer "to order." In general, however, although chemists and metallurgists can refer to a vast body of chemical knowledge, it is still difficult to predict the way different elements will react with each other in response to relatively small changes or variables: heat, proportions of particular elements, the addition of a particular contaminant, or the application of pressure.

The properties of a material, its color, electrical conductivity, and strength, for example, are dependent on the way atoms combine, and the potential for different kinds of combinations is almost endless. The application of heat and small amounts of certain chemicals can greatly increase the strength of steel, but it can be equally fragile if it contains microscopic defects. The difference between colorless aluminum oxide and a valuable ruby consists of the addition of a small proportion of chromium atoms. "Doping" a pure silicon crystal with one atom in a million of boron or phosphorous changes a poor conductor of electricity into the semiconducting material that has transformed our lives. Chemical compounds can also have the same atomic links but a different three-dimensional structure. One of the most important differences, especially in biologically active materials, is their "handedness," or what chemists call *chirality*. This refers to molecules that are composed of exactly the same atoms, but are mirror images of each other in the three-dimensional arrangement of the atoms. In biology, this characteristic determines the biological activity of the substance. One structure could be useful, the other useless or even dangerous. In living creatures, amino acids, which are the building blocks of proteins, are left-handed, while sugars are right-handed. When chemists synthesize amino acids in the laboratory, the product is usually a mixture of left- and right-handed forms. There is a growing industry of chemical suppliers dedicated to providing enzymes and catalysts to help chemists create compounds with the desired chirality.

Even the properties of the same element can vary widely depending upon its molecular and crystal structure, and upon what kind of heat and pressure it has been subjected to. Carbon, for example, is soft and slippery when it consists of sheets of hexagonal rings of atoms, as in graphite. In the form of a tightly locked lattice of tetrahedra, a three-dimensional crystal in which each carbon atom is joined to four other carbon atoms by strong chemical links, carbon becomes diamond, one of the hardest substances known. The process of concocting new compounds is frequently compared to baking (especially because it usually includes heat), in which adding extra flour to a recipe can turn a pie crust into bread, or the application of too much heat can transform a delicious dish into a lump of carbon.

As any math student knows, the permutations and combinations of even a small number of variables can quickly generate vast numbers. Thus, up to the present, the sheer volume and complexity of information has placed the ideal of a complex designed material out of reach. However, ongoing developments in computer databases and the ability of computers to crunch the myriad possible combinations and permutations of various elements is now enabling materials researchers to make significant progress. In the future, a materials researcher will be able to input a description of a desired material into a computer program whose database contains relevant knowledge about classes of materials. The computer will be programmed to make calculations that can predict new compounds and reactions and will be able to output an optimum recipe for making the material. Using computers as an aid to materials design will also cut development costs by shortening the expensive and time-consuming process of testing prototypes.

The developing field of generating computer models for materials research has been dubbed "computational materials research," and it shares similarities with other endeavors that utilize the massive computational power of computers to process large amounts of complex information. Researchers use computers, rather than their workbenches, to perform experiments and try out the myriad variations possible when large numbers of related compounds are synthesized. The resulting database of knowledge is used in the growing field of combinatorial chemistry, in which chemists synthesize thousands of compounds that have the same constituents but a slightly different atomic structure, and then test their properties using robots. Scientists are using combinatorial methods to try to solve some of the current most important problems in chemistry and materials research, such as developing high temperature superconductors and photovoltaic materials that can be used to efficiently harness the energy of the sun to generate electricity. The methods of combinatorial chemistry are also changing pharmaceutical research, where the search for new biologically active compounds continues apace (see chapter 6).

## TOOLS OF THE TRADE: NEW METHODS AID ATOMIC-SCALE MATERIALS DESIGN

Materials science is increasingly the study of the interaction of crystals, molecules, and atoms at levels where the conventional laws of physics are overtaken by the strange quantum behavior of subatomic particles. At the basis of this study are instruments that can precisely identify the atomic components of a material and the nature of the bonds between its atoms. Among the most important methods of teasing out the atomic composition of a substance are spectrometry, X-ray diffraction, and nuclear magnetic resonance, or NMR.

Spectroscopy dates back to the nineteenth century, when scientists first discovered that they could determine the chemical composition of a sample from the colors of light emitted from or absorbed by the substance. Since then, a wide variety of spectroscopic instruments have been developed. A mass spectrometer, for example, determines the chemical composition of a molecule by detecting tiny differences in the weight of its constituent atoms. Nuclear magnetic resonance spectroscopy probes the molecular structure of a material by measuring the amount of time the protons in the atoms making up the material take to return to equilibrium after being subjected to a radio signal (see page 171). In laser spectroscopy, laser pulses probe the events that occur as atoms or molecules of various elements react with each other in speeds as lightning-fast as a femtosecond—0.000000000000001 of a second. In their quest to develop harder, stronger, and lighter materials that have a uniform composition, materials scientists make frequent use of X-ray crystallography, first developed by William and Henry Bragg in the first decades of this century. This method of visualizing crystals determines the structure of the atoms in the crystal by bombarding the material with X rays. The ways in which the crystal structures refract, or bend the rays, form a distinct pattern.

Microscopy that can visualize substances at the atomic and molecular level also plays an important role. Methods of atomic-scale microscopy include atomic force and electron beam microscopy. Among the important instruments is the *scanning tunneling microscope* (STM), which was developed to harness the quantum behavior of individual electrons. The STM allows scientists to visualize the surface appearance of the individual atoms of a material by using algorithms—or special mathematical formulas—to convert the tiny movements of the microscope tip, as it ranges over the material being studied, into a "picture" of the atomic surface. (Scientists have also used the tip of an STM to arrange individual atoms, most famously to spell out the letters "IBM" in xenon atoms on top of a surface of crystalline nickel.)

## The State of the Art: The New Generation of Materials

Although the era of precision-designed materials has not yet arrived, a new generation of materials is already beginning to take a place in our everyday lives. These include smart materials, advanced composite materials, quasicrystals, and advanced ceramics.

## Smart Materials

Smart materials research focuses on substances that can be engineered to react to the kinds of stimuli that bring about a response in a living organism. These include light, vibrations, sound waves, pressure, a magnetic field, or an electrical current. Smart materials do not exhibit the same kinds of reactivity as a living organism, but they do have properties that change in response to specific stimuli. These kinds of changes can be harnessed so that the material functions as an actuator and exerts a force by changing its shape, stiffness, position, or another mechanical characteristic. This generation of materials is starting to find a place in our daily lives—in a range of surprisingly everyday devices like exercise equipment, barbecue grills, and high-end downhill skis. Smart structural beams have been developed that can change their stiffness depending on the load or the kinds of vibrations they are subjected to. Materials researchers have made rope that changes color in response to various strains, and fabrics that change their appearance in response to the surrounding landscape and weather conditions. Although the field is in its infancy, progress in smart materials research is expected to play a crucial role in the further development of robots and artificial limbs.

For example, a new material called an *electrorheological fluid* changes state from liquid to solid when an electric current is passed through it. This happens because the fluids contain a suspension of minute particles that respond to an electrical current. When a voltage is applied, the particles line up in columns, making the liquid "thicker." This kind of liquid is used to vary the response of some exercise equipment: the more "solid" the fluid, the higher the resistance of the equipment. These fluids could also be used to manufacture car shock absorbers. Depending on the kind of terrain, the shock absorber can provide more or less "bounce" in response to an electric current.

Other smart materials include piezoelectric ceramics, shape-memory alloys, and magnetorestrictive materials. Currently, the smart materials that have already found their way into the widest range of devices are piezoelectric ceramics, which can transform a mechanical stimulus into an electric charge or convert an electric charge into a force. An ultrasound machine, for example, uses piezoelectric materials in its transducer which converts an electric charge into sound waves that are used to obtain images of internal organs. Piezoelectrics are not a brand-new advanced material: Marie and Pierre Curie, who discovered radium, first experimented with piezoelectrics in 1880. Applied piezoelectrics, however, took off only in recent decades in the United States through government-funded defense and space research. Piezoelectric materials have several disadvantages: they tend to be brittle and heavy, and the piezoelectric effect disappears above certain temperatures. They also do not respond with a lot of movement, which makes them less useful than other materials where a mechanical response is needed. The advantage of piezoelectrics, however, is that they do respond quickly, and their action is already well understood.

Piezoelectric materials are now being integrated into products utilizing laminated materials. Smart packs or patches—integrated devices that include the vital components necessary for smart applications—have become a commodity item. For example, several companies manufacture piezoelectric actuator-sensor packages with preattached leads that manufacturers can easily integrate into their products. The manufacturer of a popular range of skis uses a smart pack in a ski to improve control on fast downhill runs. The energy of the vibrations is converted into electricity, which is then discarded using a "shunt circuit." This energy conversion dampens vibrations and makes high-speed turns easier. Other applications of piezoelectrics include optical tracking devices, computer keyboards, high-frequency stereo speakers, the transducers used in ultrasonography, and igniters for gas grills. The opposite mechanism, in which the application of electricity produces a change in the shape of a material, is called *electrostriction*.

Another range of smart materials with intriguing properties are shape memory alloys, which, at a certain temperature, revert back to their original shape after being subjected to a stress. In the process of returning to their "remembered" shape, the alloys can generate a large force that can be harnessed to perform a useful task. These alloys are mixtures of nickel, titanium, and copper. Such metals are already used in orthodontics: the wire in dental braces is bent to a certain shape before being inserted into the bracket. As it returns to that shape, the metal exerts a force that straightens teeth. In the future, these alloys could be used to make self-healing car bodywork. The most prominent shape-memory metals are a family of nickel-titanium alloys, a material known as Nitinol. In Japan, engineers have successfully used Nitinol to construct micromanipulators and robotic actuators that mimic the motion of a human muscle. The force is so controlled that the device can actually hold a paper cup filled with water without spilling the water or crushing the cup.

In the future, smart materials are likely to be part of systems in which embedded sensors provide information to actuators that then generate a response. Advances in micromechanical systems have created a wealth of promising electromechanical devices that can serve as sensors (see page 214). The types of smart materials likely to be incorporated in intelligent systems include optical fibers and piezoelectric materials. Piezoelectric polymers, such as polyvinylidene fluoride (PVDF), may turn up in artificial skin, because they can be formed into thin films that can be attached to a variety of surfaces. In some early experiments, PDVF films were successfully used to distinguish between different grades of sandpaper. Optical fibers could be used to provide a steady light signal to a sensor. Breaks in the light beam would indicate that a structural flaw has snapped the fiber, in the same way that elevator door controls sense the presence of a person when a light beam is blocked. Fiber sensors would also measure the intensity of light, magnetic fields, vibrations, and acceleration.

Although smart materials are already finding their way into skis and gas grills, smart structural materials will probably require extensive testing before they are incorporated into the conservative world of the construction industry. Without smart materials, the engineers who build skyscrapers and bridges tend to ensure structural integrity by overcompensating—using ten tons of concrete when five would probably have been sufficient. Replacing brawn with intelligence will enable engineers to realize considerable materials savings. In the future, smart materials embedded into a structure will be able to respond to physical forces impinging upon it by changing their color, shape, or stiffness. Thus equipped, buildings, roads, and bridges will be able to sense and possibly even automatically compensate for corrosion, metal fatigue, and age. Craig A. Rogers, Dean of Engineering at South Carolina State University, has predicted that we will soon be able to "ask" structures how they feel, where they hurt, or if they have been abused recently. In the distant future, the structure may even be able to identify the cause of the damage and repair itself.

## Advanced Composite Materials

Composite materials are truly an embodiment of the adage that a whole is greater than the sum of its parts. Composite materials are made up of two or more components that have different types of strength or resilience and, combined together, usually result in a material that is stronger than its components, and may also have other attributes—ease of molding, for example, or light weight—that make it even more useful. Composite materials have been a feature of our everyday life for millennia: bricks and paper are essentially composite materials. A more recent embodiment is fiberglass, which is used to build boats and other lightweight and strong items that will not rust or rot in water. Fiberglass is made up of strands of spun glass that have high tensile strength and can withstand a lot of stretching without breaking. On their own, these glass fibers are flexible, but locked in the resin, they resist both stretching and pressure. Similarly, resin is brittle and cracks easily, but when part of a composite it becomes tough enough to be made into automobile body panels and boat hulls.

Less familiar, and far more expensive, is a new generation of advanced composites that owe their development to government-funded military research and space programs in which high performance is considered more of an issue than cost. These new materials are classified by their composition. They include polymer matrix composites (PMCs), ceramic matrix composites (CMCs), and metal matrix composites (MMCs). These new composites, which can be found in advanced military aircraft, have advantages like high stiffness and light weight, and they have enabled the development of a wide range of products that would not be feasible with conventional materials. Since their development, advanced composites have turned up in such important aeronautic applications as the Space Shuttle; they are also to be found in a range of high-priced sporting goods, including tennis rackets, golf clubs, and bicycles. In addition to having impressive physical and mechanical properties, these materials can be tailored for a particular application. They can be reinforced, for example, at specific locations where they are most needed.

## Polymer Matrix Composites

These strong and costly composites owe their development to a new generation of "superfibers" that consist of finely spun threads of carbon or boron, for instance, which form long chains of molecules with very strong bonds between them. Because they are often smooth, even on a molecular scale, they are very strong and stiff, which are vital properties in aerospace applications. Currently, carbon fibers are the most widely used, not only because of their properties, but because they are relatively economical to manufacture. Some pundits predict that in the future, carbon fibers may completely replace metals in aircraft and other high-technology structures. Among the polymer superfibers are aramids, which are exceptionally stiff. The best known of these fibers is Kevlar, used for tires, belting, sails, and bulletproof vests.

Superfibers are incorporated into materials usually through techniques similar to those used to manufacture fiberglass. Many continuous long strands of fiber are pulled from spools and aligned. The fibers are treated and then drawn through a bath of resin, which coats them. The sticky fibers are then pressed onto a backing tape, and the reinforced sheet is rolled up. This sheet, however, is strong only in the direction that the fibers run. To make tough materials called laminates, several sheets are layered on top of one another, with the fibers running at forty-five-degree angles in successive sheets. Such laminates can be built up into complex shapes that can be cured by heat into a completely rigid structure that is stronger and lighter than a metal version of the same part.

These materials also have the advantage of being corrosion- and fatigue-resistant, and may be used as construction materials in the future if they become less costly. *Polymer matrix composites* (PMCs) may also find a welcoming home in the car industry, which is trying to reconcile the goal of producing larger cars with the need for fuel efficiency. Currently, the metal body of a car has between 250 and 350 distinct parts. Using PMCs, government analysts believe, this number could be reduced to between two and ten. Because PMCs don't rust, and are much more resistant to corrosion than metals, analysts have estimated that a PMC car body could last twice as long as the current average vehicle. Before these advanced composite materials find their way into cars, however, materials scientists will need to develop a much broader body of knowledge about their internal structure, mechanical properties, and potential failure mechanisms.

## Cermets

The hallmark of advanced composite materials is that they can be "tailored," that is, built up from constituents to have the properties required for a given application. A composite structure can also be designed to have different properties in different directions or locations. Although ceramics, which encompass a range of materials from clay to tungsten, are very useful, they are also brittle. One method strengthens ceramics by putting high-strength crystal "whiskers" in them. Another approach is to apply a thin ceramic coating to a metal substrate, which creates a component with the surface properties of a ceramic combined with the toughness of the metal.

Cermet, a material made from boron carbide mixed with aluminum or other metals, is a composite with promising qualities. When the constituents of the cermet are heated up to 1,500°F, the molten metal infiltrates the ceramic structure, forming a matrix that is lighter than aluminum but stronger than steel. In the future we could see ceramic composites used in very high-temperature applications such as advanced automotive turbine and ceramic diesel engines, heat exchangers, and other high-powered machines.

## Metal Matrix Composites

MMCs usually consist of relatively lightweight metals, such as aluminum or magnesium, reinforced with particles or fibers made of a ceramic material like silicon carbide. They are very strong, stiff, and able to withstand high temperatures. These composites once were extremely costly, and were used in military and space applications where high performance was considered more important than cost. But with the development of low-cost processing methods, MMCs are finding their way into a wide variety of applications that range from the mundane to very high technology. They include pickup truck drive shafts, brake rotors, satellites, electronic packaging, and bicycle components.

## Advanced Ceramics

Advanced ceramics are made from extremely pure microscopic powders that are subjected to pressure at high temperatures to yield a dense and durable structure. Compared with metals, advanced structural ceramics resist wear, are strong at high temperatures, and are chemically very stable. They are expected to play an important role in building construction, particularly if they are used to develop cheap and strong concrete. Another important future application for ceramics will be in medical implants: Because they are nearly inert, ceramics can be implanted in the body without toxic reactions. A ceramic with a biologically active film on its surface could form a chemical bond with surrounding tissue and encourage new tissue to form. A ceramic that can be broken down and eventually excreted from the body can provide a temporary space-filler or scaffold that will serve until the body can eventually replace it. Reabsorbable ceramics are already in use to treat jaw defects, and as composite bone plates. Polymer matrix composites are also playing an increasing role in medical devices. They have many advantages over metal implants, because they are more biologically compatible and are more reliable. In the future it may be possible to create implants from biodegradable PMC systems that would stabilize a fracture but be gradually broken down over time as the natural tissue repairs itself.

## Quasicrystals

Discovered in 1982, quasicrystals are starting to make the transition from academia to practical applications. These alloys, which combine such metals as aluminum and manganese, have fascinated materials scientists because they occupy a middle ground between amorphous and crystalline materials. The atoms in an amorphous material, like glass, are arranged randomly; the atoms in a crystal form a lattice of orderly and repeated patterns. Although the atoms of a quasicrystal are arranged in an orderly pattern, their crystal pattern is not regularly repeated. Quasicrystals differ from metals because they conduct heat and electricity poorly. Materials scientists find them exciting, however, because they "age" well; that is, they can be treated to become exceptionally hard, with surfaces that resist wear, oxidation, and corrosion. The first commercial application of quasicrystals is in a steel alloy that is made into highly precise and reliable medical instruments. Quasicrystals meet this need because they are easy to form into delicate instruments, but then can be aged for long periods, a process that makes them extremely hard and thus unlikely to break during use.

Future applications are likely to exploit properties of quasicrystals other than their mechanical ones. One possible use of titanium-based quasicrystals will be for hydrogen storage, because the atoms have an affinity for the kinds of crystal structures found in this material. Hydrogen gas is an important source of energy, because it is a clean-burning fuel. It is very flammable, however, and must be stored in a way that prevents it from exploding until it is used. Scientists have found that titanium-zirconium-nickel quasicrystals can absorb nearly two hydrogen atoms per metal atom.

# ON SILICON AND BEYOND

It is ironic that silicon, currently the most important element in the microchip revolution, becomes flint when combined with oxygen, and thus was the basis of one of the most coveted materials of the early toolmakers. The sharp edge of a broken piece of flint made it an attractive candidate for the first tools. Silicon has other important advantages: it is abundant and inexpensive, stable and nontoxic. It is usually utilized as silica, or silicon dioxide. Silica also has an important place in the communications revolution, in the form of optical fibers that can relay data at a rate of billions of bits per second. These optical fibers were first developed by researchers in the 1970s, who evaporated pure silica into a rod that they could pull into thin fibers. To ensure that light did not leak out of the fiber, the researchers developed an inner core of slightly impure silica surrounded by a cladding made of a purer glass. This made the outer cladding a reflective sheath that would send any light hitting it back into the core of the fiber.

Silicon is harder than nickel, and has the tensile strength of steel. But building a silicon chip requires painstaking, labor-intensive fabrication techniques. Thus, developing new materials that can improve on the already impressive qualities of the silicon microchip occupies the energies of many materials scientists. Already, in some applications, the silicon of computer chips has been replaced with more costly gallium arsenide, which conducts electricity faster. Researchers have already made significant progress in producing high-speed transistors from thin-film organic plastics, which could be used for "smart cards" and other cheap, throwaway memory devices, or in the array of switches that control the light emission from each picture element in conventional laptop computer displays. First developed in the mid-1980s, organic transistors had the disadvantage of being slow. There were also concerns about the stability of conducting polymers at relatively normal temperatures. In addition, many of the dopants used to impart a greater conductivity were highly toxic. Recently, however, researchers have unveiled thin-film transistors (TFTs) made from a crystalline organic film called pentacene that can almost match the speed of a silicon device. Investigators have also been working on putting thin films of light-emitting plastics and other organic materials on displays to create large, flexible, inexpensive, and efficient screens. These screens could be used for everything from lightweight backlights for computer displays to television "screens" that can be hung flat on the wall and rolled up when not required.

This is more hydrogen than is absorbed by the hydrogen-storage materials currently in use, such as the lanthanum-nickel compounds in renewable batteries in laptop computers. Another application might be in solar energy, because some quasicrystals absorb heat but do not release energy readily, so that, like the interior of a car, a quasicrystal inside a "sandwich" of other elements will get much hotter than its exterior.

## The Long Road to Applications: Fulfilling the Promise of New Materials

The public perception of modern technology is that change and innovation proceed at a breakneck pace, with new discoveries in physics, medicine, and chemistry translating almost immediately into products and applications that will further revolutionize our daily lives. The reality is that it usually takes years if not decades for an important scientific discovery to translate into a technology or application. It took thirty years, for example, from the time that the first laser was developed for this coherent light beam to become a standard feature of supermarket checkout counters. This lag time is especially pronounced in the case of some newly discovered materials that have alluring properties, but have yet to make their way into manufactured products. Among those materials are high-temperature superconductors and fullerenes—widely known as "buckyballs."

## High-Temperature Superconductors

In recent decades, advanced physics has become increasingly abstract; physicists use supercolliders to conduct esoteric research deep into the heart of the atom, where particles live and die in a trillionth of a second. A profound scientific phenomenon that everyone can grasp, on the other hand, is superconductivity—the discovery of materials that, when cooled, conduct an electrical current with no resistance. Related to this is magnetic levitation, a phenomenon that a high school science student can demonstrate. Liquid nitrogen, which boils at 77 Kelvin, or −321°F, can be stored for several days in an ordinary thermos bottle. When it is poured over a disk of ceramic superconductor in a dish, a magnet placed above the disk floats in the air, repelled by the magnetic field of the superconductor.

Kits to demonstrate this phenomenon can be ordered from a science fair catalog. And, if it is not available for overnight delivery, the superconducting material is relatively easy to synthesize: take one part of yttrium, two parts of barium, and three parts of copper and grind up the materials using a mortar and pestle. Put the powder in a heat-resistant dish and bake in an oven at 900°C for a few hours. Grind up the powder and place it in a metal die. Squeeze it until it becomes a circular disk. Place the disk in an oven through which a steady stream of oxygen flows, and bake again. After this disk is slowly cooled, it will be a superconductor.

Superconductivity was first discovered in 1911, when a Dutch physicist, Heike Kamerlingh Onnes, chilled mercury with liquid helium to four degrees above absolute zero, or 4 Kelvin, and found that the mercury made a sudden transition: it would transmit electricity without loss. (Zero degrees Kelvin is equivalent to −273°C or −460°F. It is considered the point of absolute zero, a temperature at which, in principle, all atomic motion ceases and atomic particles have no kinetic energy.) The phenomenon of superconductivity is explained by quantum theory describing the behavior of electrons. In its simplest form, this theory proposes that at the low temperatures at which certain materials make a transition to a superconducting state, the electrons condense into a "quantum ground state" which permits electrons that normally repel each other to pair up and pass through the medium unobstructed. These electrons are called *Cooper pairs*.

Under normal circumstances, even the best electrical conductors, such as copper, possess resistance to the flow of electrons, transforming a portion of the electrical energy into heat. The energy lost in conducting electricity is proportional to this resistance, and the possibility of transmitting energy in a superconductor, with zero resistance and total efficiency, has important implications. If electric utilities could generate and transmit current without loss, there would be great environmental and economical advantages because less fuel would have to be burned to generate electricity, thus cutting down on pollution and the use of valuable fuels. The electric circuits of devices like computers would function at lightning-fast speeds because the electrons in the circuits would encounter no resistance. Superconductivity would also be a useful way to store electricity: an electrical current would keep on going around in a superconducting loop without energy loss until it was needed.

Superconductivity is probably the closest engineers have come to their dream of perpetual motion, in which a machine set in motion is not hobbled by the forces of friction and inertia. But this tantalizing vision is hard to realize if materials have to be cooled by costly liquid helium to the icy temperatures of deep space before they become superconductors. In the 1980s, however, researchers discovered substances that make the transition to superconductivity at temperatures that can be attained using liquid nitrogen, which, unlike helium, is abundant and cheap. These materials have been dubbed high-temperature superconductors. The temperatures at which they make the transition to superconductivity are still extremely low, but are more easily attainable using affordable means. One class of these materials are ceramiclike substances known as *cuprates*; the first to be discovered was lanthanum barium copper oxide, which begins to super-conduct at 35 Kelvin. This was the forerunner of an even more exciting find: the discovery of yttrium barium copper oxide, or YBCO, which makes the transition to superconductivity at 93K. More high-temperature superconductors have since been discovered. In 1994 a new copper oxide superconductor was synthesized that had zero resistance at 133K.

Although these discoveries have caused enormous excitement in the physics community, high-temperature superconductors have so far had a small impact because they carry only a limited amount of electricity be-fore they lose their superconductivity. They are also brittle, and hard to draw into wires, which would be a requirement for electricity transmis-sion. There are many hurdles that must be overcome before large-scale and transforming applications of high-temperature superconductivity become a reality. One challenge is to find a way to form these brittle sub-stances into wires, which would be a requirement for electricity transmis-sion. Another is to develop high-temperature superconductors that can carry large loads of current without reverting to a normal state. Methods of forming superconducting wires have included placing the substances into silver tubes that are then compressed and surrounded by nitrogen coolants. Another more promising approach is to lay a precisely aligned thin film of superconducting material on top of a substrate using methods like molecular beam epitaxy and chemical vapor deposition.

Companies that build transmission lines for electrical utilities are conducting serious research into fabricating cables with high-temperature superconductors. Prototype cables that range from three to about one hundred feet long have already been built, but many technical challenges must still be overcome. These include developing practical methods of splicing a cable cooled to $-321°F$ into a system operating at ambient temperatures. The cable must also be shielded from lightning and other conditions that could disrupt the current.

As thin films, high-temperature superconductors are finding their way into practical uses, as supersensitive magnetic detectors used for medical diagnosis, and in very narrow-waveband microwave filters for communica-tions. Such magnetic detectors have been dubbed SQUIDs—or *supercon-ducting quantum interference devices*. Because of their superconducting state, they are capable of registering very small changes in a nearby magnetic field, and are presently the most sensitive magnetic detectors available. Tests have shown that SQUIDs can pinpoint the areas of the brain respon-sible for the electrical disturbances of focal epilepsy, which means that they can detect a magnetic field a billion times weaker than that of the Earth.

There are also predictions that SQUIDs will become standard issue for testing such structures as oil pipes and bridges, because fatigued metal produces a unique magnetic signature. SQUIDs are manufactured using both conventional low-temperature conducting materials and newer *high-temperature superconducting* (HTS) materials. As HTS technology improves, it will be utilized more extensively because of cost and safety advantages.

Microwave technologies are another important major application of thin-film high-temperature superconductors. Sometimes called radar waves, microwaves are electromagnetic waves that oscillate at very high frequencies (from 3 billion to 300 billion cycles per second). Because of the energy state of their electrons, materials in a superconducting state do not absorb microwave energy, and thus provide a very good shield—or filter—for electromagnetic radiation. These filters are used in cellular telephone communications. Applications are also being developed in which thin films of high-temperature superconductors are used to shield sensitive electrical components from electromagnetic radiation. (In fact, to protect astronauts or other living things from electromagnetic radiation in space, scientists have proposed applying a superconducting film to the outside of spaceships.) Superconductors can also be used to build a device called a delay line, which stores microwave energy for a short period of time, simply by providing an extra-long path for microwaves to traverse.

The magnetic properties of superconductors are also an important element of their utility. When an electric current passes through a wire, it generates a magnetic field—the bigger the current, the bigger the field. The phenomenon, called Ampere's Law, makes possible electric motors and most other devices that convert electrical energy to mechanical work. (Strong magnetic fields are also used in advanced medical scanners and electron microscopes.) The problem with electromagnets is that the energy lost because of electrical resistance increases rapidly as the current is increased. That lost energy appears as heat, which, in the case of very large magnets, must be removed somehow before the magnet melts or burns. A superconducting electromagnet, on the other hand, can generate very powerful magnetic fields without overheating, and without being very large. Particle physicists were among the first to employ superconducting magnets, which warp the trajectories of speeding charged particles. These magnets are key ingredients of most of the detectors used to analyze subnuclear collisions. The powerful Tevatron accelerator, located at Fermilab in Illinois, is a ring of superconducting magnets 3.9 miles in circumference.

The ability of superconductors to generate large magnetic fields has led to the development of one of the few commercially successful applications of low-temperature superconductivity, the magnetic resonance imaging or MRI machine, which uses the fields generated by a low-temperature superconductor to image soft tissue inside a person's body that cannot be seen by other means (see chapter 7). Other applications of magnets constructed from classic low-temperature superconductors are high-field magnets and the magnetic field measuring devices called magnetometers. As the level of technology advances, other instruments fabricated with new high-temperature superconductors will eventually replace these devices. Because of the higher transition temperatures of the new materials, the devices would be cheaper both to manufacture and operate.

Another magnetic feature of superconductors that probably inspires technological visionaries the most is the phenomenon of magnetic levitation. This dramatic property of superconductors is called the Meissner effect, after the German physicist Alexander Meissner, who discovered it in 1933. He found that superconductors are magnetic shields—they cannot be penetrated by another magnetic field. As a result, a superconductor will be repelled by a magnetic field in its vicinity and tend to move away from it. The mutual repulsion of a magnet and a superconductor is responsible for the phenomenon of levitation, and has its embodiment in the dream of a magnetically levitated train (or maglev). In some maglevs, superconducting magnets are located on the bottom of the train, and the tracks are a series of closed wire loops aligned down the path of the train. Proponents dream of a high-speed train that can travel on a specially constructed monorail at speeds of three hundred miles per hour or more because it is not slowed down by friction. Although prototype maglev superconducting trains have been constructed in Japan, enormous infrastructure requirements, and the continuing uncertainties of the technology, has quelled much of the interest elsewhere.

## Josephson Junctions: A Key to Warp-Speed Switching

An unusual property of superconductors, known as the Josephson effect, forms the basis for a wide range of electronic applications. Named for Brian Josephson, the Cambridge University physicist who predicted its existence, the Josephson junction consists of two superconductors separated by a thin insulating strip. The strange laws of quantum mechanics predict that the superconducting pairs of electrons called *Cooper pairs* will tunnel through this strip when a current is applied. SQUIDs, the highly sensitive detectors of magnetic fields, consist of two Josephson junctions connected in parallel. Josephson junctions can also be used as ultrafast electronic switches from which it might be possible to build powerful tiny computers. Superconducting switches made from Josephson junctions dissipate hardly any heat at all, and their switching times are measured in picoseconds, or trillionths of a second. With such advantages, one may wonder why a superconducting computer has not appeared on the horizon. The answer so far is that the mass production of Josephson junctions has not proved economical, nor has the need to cool the system with liquid helium. American efforts to build such a computer were dropped in 1983, but efforts by a Japanese consortium have continued.

## A Revolution in the Making:
## Buckyballs and the Development of Fullerene Chemistry

Of all the 112 known elements, carbon can form more compounds than any other. The reason is that carbon has a remarkable ability to form bonds in which it shares pairs of electrons with other carbon atoms (these are known as covalent bonds) in an apparently infinite range of combinations. The enormous field of organic chemistry is devoted to the study of carbon compounds. (Scientists dubbed it organic chemistry because they initially believed that carbon compounds could only be made by living organisms.) Today, more than 8 million synthetic and natural organic compounds have been discovered or synthesized. It is little wonder, then, that materials scientists are so fascinated by buckminsterfullerene, or $C_{60}$, a molecule that consists of sixty atoms of carbon joined together in a hitherto unexpected spherical configuration. Since the 1985 discovery of buckminsterfullerene, many other related molecules have been discovered or made: they all belong to a class of substances called fullerenes. Named "Molecule of the Year" by the journal *Science* in 1991, buckminsterfullerenes—quickly thereafter dubbed buckyballs—are expected to lead to the development of a brand-new field of "sphere" chemistry and the production of novel substances with chemical and physical properties never before seen. Future applications of fullerenes could range from drug delivery and superconductivity to nanotechnology (the conception and development of machines and devices whose size is measured in molecular dimensions).

Before buckyballs were discovered, carbon was thought to exist in only two naturally occurring crystalline forms: diamond and graphite. Although both are essentially pure carbon, diamond and graphite are so different from each other because of the arrangement of the carbon atoms within their molecular structures. These two different structures result in radically different appearances and properties: diamond is the hardest known material, while graphite is darker in color and much softer. Graphite is a useful lubricant and electrical conductor; diamonds make wonderful drill bits and do not conduct electricity.

Fullerenes have a completely different structure from either diamond or graphite; they are spherical and tubular, shapes that scientists believe will endow these exotic molecules with exciting and novel properties. Fullerene chemistry has its origin in the study of interstellar dust: the scientists who discovered $C_{60}$ were attempting to simulate the conditions that would result in the formation of carbon dust around and in between red giant stars. To do so, they vaporized graphite at 10,000°C using a powerful laser beam. Then they examined the soot produced by this method and found that most of the resulting carbon molecules contained sixty atoms. Scientists then started to figure out how sixty carbon atoms could be arranged to produce a stable molecule and arrived at a structure consisting of twenty hexagons and twelve pentagons: the sixty-carbon atom molecule in the soot obtained from vaporized graphite was basically a molecular-sized soccer ball! Because of their resemblance to the geodesic domes erected by the American architect and engineer Richard Buckminster Fuller, scientists dubbed the carbon spheres *buckminsterfullerenes*. As it happens, $C_{60}$ is the smallest closed structure that can be made in which twelve pentagons are isolated from each other by hexagons. As the number of carbon atoms is increased, the next similarly closed structure is $C_{70}$. A host of fullerenes have since been produced, including closed spheroids containing 540 carbon atoms, and "onions," consisting of concentric layers of buckyballs. Hollow tubes (dubbed *buckytubes*, or nanotubes), which are among the smallest electrical "wires" ever made, have also been created. These include nanotubes

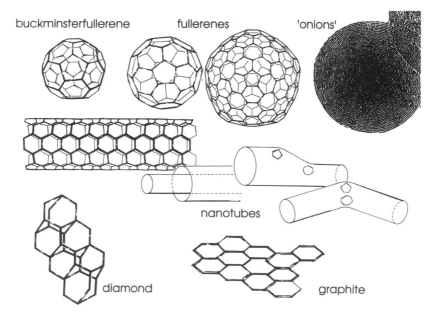

**Figure 8.1**   Carbon in the form of fullerenes may become the components of microscopic machines.

made of a single layer of carbon atoms, and multiwalled nanotubes that consist of layers of tubes wrapped around each other.

Since their discovery, the uniquely spherical structure of fullerenes, coupled with the propensity of carbon to form compounds, has tantalized materials scientists with its potential. Molecules can be made, for example, in which metal atoms and other chemicals are "trapped" inside the buckyball. In the future it may be possible to insert a drug inside the "soccer ball" and create "doors" on the surface of the sphere that open when the drug needs to be released at the right time and place in the body. Other substances have been made in which chemical groups are added to the outside of the soccer ball molecule, creating ears—or "bunnyballs." Because one particular kind of molecular structure resembles that of the low-friction coating material Teflon, scientists believe that buckyballs can be utilized as superlubricants. The slippery balls, they presume, will roll past each other with great ease. Other applications include the development of rocket fuel, an application that is based on the discovery that spherical fullerenes truly behave like balls: in one experiment, buckyballs fired at walls of graphite and silicon at 17,000 miles an hour bounced right back. Ejecting the accelerated buckyballs would produce the forward thrust of the rocket engine in the same way as ejected gases drive a jet engine forward.

The molecular structure of fullerenes—and their tendency to self-assemble—has also proved fertile theoretical grist for nanotechnologists who envision using fullerene nanotubes and shells bonded with chemical compounds or "doped" with metal contaminants to build nano-scaled devices many times smaller than the silicon-based semiconductor devices that already exist.

But for all this, spherical fullerenes, like high-temperature superconductors, have so far failed to reach the potential that materials scientists firmly believe they possess. The reasons for this failure are instructive of the kinds of hurdles materials scientists face when developing new materials. For example, it was found that while fullerenes are indeed a superior lubricant, they evaporate at 300°C, which is far too low a temperature for them to be useful. Engines that need to be lubricated operate at far higher temperatures than this. Similarly, although fullerenes exhibit promising high-temperature superconductivity when doped with elements like potassium and rubidium, the doped fullerenes degrade rapidly in air. And at this point fullerenes are simply too costly to be used as raw material for the construction of affordable devices. In the future, however, fullerenes may find a way into useful applications that materials scientists have yet to dream of. Meanwhile, although finding applications for buckyballs continues to elude researchers, they continue to be optimistic about the potential of nanotubes. These minute graphite tubes are just a few nanometers in diameter and about 100 microns long. They are flexible, lightweight, and extremely strong, and are good electrical conductors and heat transporters as well. Among the possible applications are wires for atomic-sized electronic devices and superstrong cables.

## Micromechanics: Materials Science Meets Electronics

While fullerenes could potentially be the raw material for the micromachines of the twenty-first century, materials science, electronics and mechanics are already converging to craft a range of mechanical devices that are smaller than the eye can see. They are accomplishing this using the same chip fabrication methods that enable a silicon microchip to contain millions of microscopic electrical circuits in an area the size of a postage stamp.

The process of crafting microscopic devices—such as gears, sensors, and tiny motors—often using the same materials and processes as the microchip industry, is called *micromechanics*. And, like transistors, millions of these devices can be manufactured at the same time. By combining electronic circuits and microscopic mechanical structures, engineers can equip electronic systems with physical and mechanical capabilities that supplement the processing and storage capabilities of a microcircuit. Miniature systems that integrate mechanical and electrical components have been dubbed *microelectromechanical systems*, or MEMs, and the field is already producing devices that are at work in our daily lives. Thermal ink-jet printers use micromachined devices to dispense ink. In some automobiles a tiny acceleration sensor, fabricated on the surface of a silicon microchip, triggers the release of an air bag when a car crashes. This sensor changes the position of tiny suspended beams when a car decelerates rapidly. The resulting change in the amount of electrical charge in the device deploys the air bag. In development are micromachines able to detect and control motion, light, sound, heat, and other physical forces. Thus, in the future, micromechanics combined with microelectronics could result in the development of a sensor that can be placed at the tip of a hypodermic needle. The sensor could detect the minute quantities of a particular drug or hormone in the bloodstream, or could be used in an implant to dispense important hormones like insulin into the circulatory system. These sensors could also be incorporated into artificial skin that can sense differences in temperature and pressure. The added advantage of these tiny machines is the sensitivity that comes from their minute dimensions. A biosensor could be used in environmental monitoring to react to the presence of tiny quantities of a hazardous chemical at levels that would probably elude a larger sensor.

One technique used to manufacture the motion sensor and other MEMs devices is known as *surface micromachining*. Beams and other microscopic structures are built from a film of silicon, a few microns thick, that is deposited upon a microchip. This technique uses many of the same methods used in electronics fabrication. Photolithography uses ultraviolet light to create a pattern on a chip, marking off an area that is etched away to leave holes where the micromechanical devices will anchor. Then another type of silicon is deposited in these holes to build up the structures. The processes of lithography, etching and the depositing of material, are repeated until a beam or a tiny motor is completed.

One way in which MEMs may have an impact on our lives is to create television images the size of a movie screen. A large semiconductor manufacturer has built a MEMs device that projected such an image using 2 million moving metal mirrors installed on a microchip. Each mirror is a picture element that is illuminated or darkened. The mirror consists of a 16-micron square of aluminum that, for an illuminated pixel, reflects pulses of colored light onto a screen. Each small picture cell—or pixel— in the screen is turned off or on when an electrical field causes the mirror to tilt 10 degrees to one side or the other. In one direction, the light beam is reflected onto the screen to illuminate the pixel. In the other, it scatters away from the screen, and the pixel remains dark. The image gets projected onto a television screen with a high degree of brightness and resolution of picture detail.

MEMs could also have a future in aerodynamic design. Currently the turning, ascent, and descent of an airplane are controlled by the

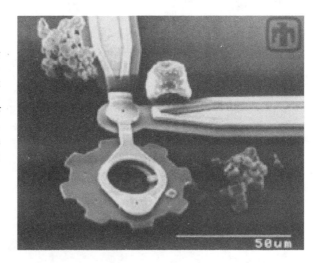

**Figure 8.2** A drive gear chain fabricated using surface micromachining at Sandia National Laboratories is shown with a grain of pollen (top right) and coagulated red blood cells (lower right and top left) for scale.

**Figure 8.3** This microscopic multiple gear speed reduction unit was fabricated at Sandia National Laboratories using methods similar to those used to lay down millions of electronic circuits on a thumbnail-sized piece of silicon.

large moving surfaces on the wing—the flaps, slats, and ailerons that are deployed to slow the airplane down during descent. In search of greater maneuverability, stability, and flying efficiency, engineers have proposed replacing these devices with a "smart skin" made up of thousands of 150-micron-long plates, or actuators, that will move in response to an electrical signal. In their resting position, these plates would remain flat on the wing surface. When an electrical voltage is applied, the plates would rise up from the surface at up to a 90-degree angle. Thus activated, the plates could control the currents of air across the wing. Sensors could monitor the current of air rushing over the wing and send a signal to adjust the position of the plates. With the additional aerodynamic control provided by these microscopic navigational aids, aircraft engineers believe it may be possible to develop radically different aircraft that will be more stable and maneuverable than current designs.

## Self-Assembly and the Machines and Materials of the Future

It appears increasingly likely that high technology will remain focused at the atomic and molecular level. The machines and devices of the future will be so small that they cannot be built with current methods; they will virtually fabricate themselves. The method by which these machines will be constructed has been called self-assembly, a procedure in which humans set up the preconditions for the process but are then not actively involved. Rather, atoms, molecules, aggregates of molecules, and components arrange themselves into ordered, working elements without human intervention.

Already, some common products are manufactured using self-assembly—or the natural tendency of various kinds of molecules to "organize" themselves. Most window glass, for example, is made by floating molten glass on a pool of molten metal. The metal tends to minimize its surface area by becoming smooth and flat, so that the glass on top becomes optically smooth and uniformly flat as well. Similarly, scientists do not at this time dictate how the silicon and dopant atoms in a semiconductor crystal are organized. Rather, their placement is dictated by the laws of physics.

Materials scientists have also exploited the principles of self-assembly to develop *liposomes*, which are tiny spherical capsules with a hollow interior. These liposomes are good for transporting drugs in the body because they protect their cargo from being degraded by enzymes. They are also being used as a method to deliver genes into human cells that prevents them from being degraded first (see chapter 6). To develop liposomes, scientists emulated the way cell membranes are formed. The membranes are made mostly of molecules called *phospholipids*, which have a kind of two-sided nature: one end of the lipid is attracted to water and the other end is repelled by it. When placed in a watery environment, the molecules spontaneously form a double layer, or bilayer, in which the hydrophobic, water-repelling ends point toward one another. Researchers use these same phospholipids to make liposomes. If there are enough molecules, the phospholipid bilayer will grow into a sphere with a cavity large enough to hold drug molecules. The liposomes are then injected into the body and the drugs are released when the sphere leaks or ruptures.

## Self-Assembling Monolayers

Some tubular molecules also have different properties on each end. This phenomenon is used to develop crystals that adhere together onto one surface and will have different properties on their other surface. These films are called *self-assembling monolayers*, or SAMs. The molecules in a SAM are sausage-shaped. At one end is an atom or group of atoms that interact strongly with the surface; at the other, chemists can attach a variety of atomic groupings, thereby altering the properties of the new surface formed by the SAM. The process can be repeated, with additional layers being built onto the monolayer. Researchers have used the self-assembling properties of these monolayers to build films that are just molecules thick. Such monolayers are being used to guide the growth of living cells, and to form microstructures that will be used in future generations of microcircuits. In addition, researchers are exploring the use of self-assembling monolayers to control the reflectivity of devices that use light in communications.

# POLYMERS: MOLECULAR GIANTS

Polymers, the giants of the chemical world, are synthetic or natural compounds that can consist of millions of repeated units of simple molecules that are linked together. Material scientists are mindful of the fact that nature provided us with polymers well before the plastics industry figured out how to synthesize them. Among polymers that come to us courtesy of nature are rope, skin, hair, DNA, and lignin, the main component of wood. The first semi-synthetic polymer was celluloid, developed as an alternative to ivory billiard balls in the 1860s. The first completely synthetic organic polymer was invented by Leo Baekeland in 1906. Bakelite does not react with solvents and is a good electrical and heat insulator. Because of these properties, it was a boon to the developing electrical industry.

Now, synthetic polymers are found everywhere in modern technological societies, tailor-made to suit particular needs. Most commercially important polymers are organic; that is, they are compounds of carbon. But some are inorganic; for example, many industrial lubricants are made of polymers based on silicon. The properties of a polymer depend on the way its constituent molecules are linked in three-dimensional space. By controlling conditions during the synthesis of a polymer, chemists can determine its size and shape, "designing" molecules with a range of useful properties like toughness, elasticity, and the ability to form fibers.

Some plastics—the ubiquitous polymers of our modern life—enjoy a very long life after they are discarded, because although they consist mainly of carbon, hydrogen, and oxygen, no bacteria have yet evolved that can make a meal of their particular molecular structure. Owing to concerns about the environment, polymer chemists are developing methods of manufacturing biodegradable polymers. One way of doing this is to incorporate sugar molecules into the polymer chain. Bacteria digest the sugar, breaking the chain into small fragments. Researchers are also using genetic engineering to produce biodegradable plastics by taking advantage of the fact that more than one hundred polymers are produced by bacteria. To increase this production, researchers have inserted polymer-manufacturing genes from these bacteria into those of *Escherichia coli* to produce large polymers. Researchers have also turned plants into plastics factories by giving them bacterial genes.

In the burgeoning field of smart materials research, scientists are looking to biological polymers to serve as design models for new high-tech polymers, particularly because of their ability to respond to environmental changes. Already, materials scientists have synthesized biomimetic materials that respond to various environmental cues such as changes in acidity, temperature, voltage, and pressure.

## Nature's Way: Biomimetic Materials Research

As materials science becomes increasingly interdisciplinary, some researchers are taking a careful look at how materials are constructed in nature, with an eye to developing methods of "growing" materials similar to those used by living organisms in growing and in fashioning intricate structures.

Materials scientists who study nature's structures express considerable admiration for the structural and mechanical engineering skill they find. Most living tissues have to carry loads, and some tissues, like muscles, not only have to apply loads as well, but must change shape while doing so. Until the recent introduction of synthetic materials, the human species relied on animal skins for tough, flexible items like leather sandals, shoes, and horses' harnesses. Biology also places a premium on strength and mechanical safety, while requiring that tissues be light in weight and metabolically efficient. Even the largest and heaviest bones of a large bird, for example, combine hollow spaces for lightness and supporting structures for strength. In emulating nature, materials scientists hope to benefit from the best of both worlds. They hope to develop advanced materials that use the best qualities of natural structures, but have the advantage over nature of being quickly fabricated (in contrast to the time that living organisms take to construct the hard parts of their bodies). Genetic engineering technologies could potentially play an important role in creating biomimetic materials. Genetically modified bacteria are already used to produce recyclable polymers in the plastics industry. Researchers are also studying the biochemistry of spider webs, which are remarkably strong and elastic.

One of the attractions of biomimetic materials research is that nature uses such readily available and inexpensive raw materials as oxygen, hydrogen, carbon dioxide, and calcium. In addition, nature builds strong materials at low temperatures using environmentally benign materials. This is in contrast to man-made materials, often constructed from toxic materials worked at high temperatures. The resulting natural structures may have impressive properties: for example, an abalone shell, constructed of chalky calcium carbonate, resists fractures almost as well as synthetic ceramics like zirconia and carbon boride. An additional feature of natural materials is that they are "self-assembled." The molecules of natural materials form spontaneously from smaller precursors, with no poking or prodding from us and our clumsy tools. Emulating the principles of self-assembly might help us to make molecules and molecular machines that are too small to be made by conventional synthetic techniques. In nature, enzymes mediate the chemical reactions that control both the creative and energy-producing life processes of a cell. These are proteins that facilitate a chemical reaction between two substances but do not themselves take part in the reaction. In chemistry, an analogous role is played by catalysts, which facilitate a chemical reaction between two substances but do not get used up in the process. Chemists predict that catalysts will probably play an important role in self-assembly.

## OUR MATERIAL WORLD: FROM POTTERY TO COMPUTER CHIPS

There are four major classes of materials: metals, polymers/elastomers, composites, and ceramics/glasses. In the ancient world, gold, copper, bronze, and iron attained wide use. The natural polymers that our ancient ancestors exploited were wood, skin, fibers, and glues. The composite materials they created were straw bricks and paper; the ceramics and glasses were stone, flint, pottery, glass, and cement. In our era, the most important metals are steel and aluminum; the most widely used polymers are synthetic plastics; the composites are materials like fiberglass and Kevlar, and the most widely used ceramic is silicon. The latter, in the form of silica, occurs abundantly as quartz, sand, flint, and agate, and is used to make glass, cement, and computer chips.

Cast iron first attained general use in the fifteenth century; steel was first developed two hundred years later, and steel alloys were developed in the nineteenth century. New materials, particularly plastics, are replacing traditional steel applications. With the invention of Bakelite in 1906, a new era of synthetic polymers, called plastics, was launched. The development of polymers that served as highly functional substitutes for natural polymers, like cotton and silk, first took off in the 1930s with the invention of nylon. Since then, numerous synthetic polymers, such as polyvinyl chloride, polystyrene, and Plexiglas, have attained wide use in industrialized societies. Metals like titanium and aluminum have found their way into sporting goods and aerospace. Aluminum is gaining an increased market share because of its light weight and corrosion resistance. As recently as the 1960s, composite materials were not considered very important, but it is projected that by 2020, all four classes of materials will have roughly equal importance.

# GLOSSARY

**actuator**   A device that activates a mechanical device or puts it into motion.

**aging**   The process by which alloys, which are mixtures of metals, harden over a period of time.

**alloy**   A substance composed of two metals that are thoroughly combined at the atomic level.

**amorphous**   Term used to denote a material that does not have a distinct crystalline structure.

**atom**   A particle that is the smallest unit of an element. The atom has all the characteristics of that element and comprises a nucleus, consisting of protons and neutrons, surrounded by orbiting electrons.

**Bose-Einstein condensation**   A state of matter in which such particles as Cooper pairs or helium atoms are reduced by supercooling to their lowest energy stage. In that state they have special properties, e.g., superconductivity or superfluidity.

**cermets**   Materials made from one ceramic and one metallic component. Cermets are exceptionally hard, and are able to withstand high temperatures.

**composite**   A material made of different elements or parts that are combined in some way. The properties of the composite are often more useful than those of its parts.

**compounds**   In chemical terms, a compound consists of the atoms or ions of two or more different elements that have formed atomic bonds that cannot be physically broken.

**compressive strength**   A measure of the strength a material has when it is put under pressure.

**conductor**   A material that conducts electricity well, i.e., a material that has low electrical resistance.

**crystal**   A structure in which atoms are stacked together in regular patterns, which repeat themselves regularly. (In amorphous materials, by contrast, the atoms are haphazardly arranged.)

**doping**   Adding a trace amount of a foreign element to the crucible in which crystals grow, to change the conductivity of the material.

**epitaxy**   The process of growing the crystals of one mineral on top of the crystal of another mineral. The method used to grow the crystals ensures that both crystals are oriented (or point) the same way. This enables materials scientists to control their properties.

**extrude**   To thrust, force, or press out a viscous liquid; to form a material to a desired cross-section by forcing it through a die.

**graphite**   A form of carbon in which the atoms are arranged in sheets with hexagonal patterns stacked one above the other.

**hole** In an atom, the absence of an electron in the outermost orbiting shell. This "shell" is also called an *electron band*; a hole carries a positive charge, and electrons move toward it.

**insulator** A poor conductor of electricity.

**ion** An electrically charged atom, such as a single fluorine atom, that is negatively charged.

**isomers** Molecules that have the same structure but differ from each other in their three-dimensional arrangement, or chirality, so that they appear to be mirror images of each other are called isomers. They are often biologically important. A molecule that is not correctly "handed" often does not exhibit the same biological activity.

**isotopes** Atoms of a particular element that have the same number of protons in their nuclei but differ from each other in their number of neutrons. Changing the number of neutrons in the nucleus converts an atom of an element into an isotope. Isotopes are often radioactive, and decay to form new elements.

**Josephson junction** Two superconductors separated by a barrier through which Cooper pairs (a pair of electrons in a superconducting state) can tunnel from one side to the other. The tunneling electrons have escaped the confines of the superconductors.

**molecules** A group of atoms that is held together by chemical forces. A molecule can consist of atoms of the same element, such as two atoms of oxygen, or of two or more different elements, such as a molecule of table salt, or sodium chloride.

**piezoelectricity** An electric charge generated by "dielectric" crystals that are subjected to a mechanical stress. This also refers to the reverse process, in which such crystals generate a force when an electrical charge is applied to them.

**polymer** Long-chain molecules formed when many individual small molecules called monomers are joined together.

**rare earth** One of fourteen metallic elements in the periodic table with atomic numbers from 57 through 70. These elements appear to have promising potential in the development of high-temperature superconductors.

**superconductivity** The flow of electrons without resistance. This phenomenon occurs in certain metals, alloys, and ceramics at temperatures near absolute zero. In high-temperature superconductors, the state of superconductivity is attained at relatively warmer temperatures.

**tempering** Hardening or strengthening metal or glass by various processes, which can include heating, or alternate heating and cooling.

**tensile strength** The measure of the maximum amount of pulling stress a material can withstand without tearing or breaking.

# 9

# Mother Earth, Wind, and Fire: Energy for a Small Planet

To a large degree, the history of technology is written in terms of innovations in energy generation and the invention of machines that efficiently, reliably, and safely harness that energy to perform work. And the level of technological advancement of a society is often measured by how much energy the society produces, and how efficiently it is used. Compare, for example, the first steam engine, built in 1712 to pump water out of the tin mines of Cornwall. This engine towered more than fifty feet high and worked at just five horsepower. By contrast, a modern jet engine can deliver enough thrust to lift a fifty-five-ton deadweight one meter in one second, and each engine of a Boeing 747 can deliver 20,000 horsepower at cruising altitudes.

Nevertheless, the lumbering steam engines that powered the railroads and mills of the Industrial Revolution and the sleek, efficient machines of our era continue to have much in common: they are powered by the combustion of fuels that have their origins in the complicated process of photosynthesis, in which plants use the energy of sunlight to turn carbon dioxide and water into food. And both emit residue gases—notably carbon dioxide—in the process. Finding ways to meet the planet's huge appetite for cheap and plentiful energy that can be generated without releasing carbon dioxide and other polluting gases into the atmosphere is one of the most important technological and political challenges facing modern society.

## Our Carbon Energy Diet

Such energy sources as coal, crude oil, and natural gas—the so-called fossil fuels—were formed hundreds of millions of years ago, when freshwater swamps and shallow seas covered vast areas of the globe. Over millions of years, dead animal and plant matter accumulated at the bottom of these seas, where it was gradually buried under layers of sediment, and converted by pressure and heat to coal, crude oil, and natural gas. Because carbon is both the feedstock for energy storage in plants and the by-product, in the form of carbon dioxide, of its combustion, fossil fuels (and firewood) are called carbon-based fuels.

Every aspect of modern life depends on the availability of large amounts of energy generated at reasonable prices, and at this time, fossil fuels are the major source of all energy worldwide. If only for economic reasons, most people appreciate the need to conserve energy and lessen the industrial world's dependence on fossil fuels. This is often presented as an issue of depletion: created over millions of years of biology and geology, these fuels cannot be renewed and will eventually run out. In addition, many countries, without natural energy resources, are forced to import a large proportion of their fuel needs. This puts them at the mercy of geopolitics and is also an enormous financial burden, particularly for developing countries.

In fact, though it is certainly true that fossil fuels will eventually be depleted, new techniques of exploration and resource extraction have led some analysts to estimate that we have enough crude oil to last for at least the next half-century, and sufficient coal reserves for another two centuries. In the United States, which leads the world in energy consumption, prices have remained so low that one of the most popular new vehicle purchases continues to be sports-utility vehicles, whose fuel consumption is close to that of the gas-guzzling, tailfinned "land yachts" of the 1950s and 1960s.

The more immediate concern, according to some scientists and most environmentalists, is the effects on the environment of our dependence on fossil fuels. Coal mining and oil-well drilling are environmentally destructive activities. Far more serious are the effects of burning fossil fuels on the Earth's atmosphere. The gases and particles produced by the combustion of carbon-based fuels are responsible for the formation of acid rain, and the release of damaging ozone and volatile organic compounds (VOCs) into the air we breathe (see chapter 10). Of most concern in this chapter is the increasing likelihood of global warming caused by the release of "greenhouse gases"—notably carbon dioxide—into the atmosphere. This topic is arousing enormous passions: If global warming has begun, and the consequence is as serious as predicted, then reversing or slowing this process will require drastic cuts in the worldwide use of fossil fuels, leading to radical changes in the production and use of energy. Such an alteration could cause global upheaval: In the developing world, rising standards of living are expected to lead to huge increases in energy requirements. Numerous industries have considerable and entrenched investments in fossil fuels, including energy companies, utilities, and car manufacturers. Rapidly industrializing countries like India and China have enormous coal reserves and every intention of using these reserves to fuel their growth. National economies in such countries as Nigeria, Indonesia, and the Arabian Gulf States depend on fossil fuel production.

## What Is the Greenhouse Effect?
## Or, Why There Aren't Any Martians . . .

If not for the greenhouse effect, the Earth would be a lot like Mars, which has an average daily temperature of $-27°F$—and there wouldn't be any Earthlings, either. The sun's energy comes through the atmosphere in light waves and is absorbed by the Earth and converted to heat energy at the planet's surface. The infrared heat energy radiates back upward through the atmosphere and into space. Mars's thin atmosphere cannot prevent heat from escaping, but the greenhouse gases and water vapor present in the troposphere—the layer of the Earth's atmosphere from the surface to about ten miles in altitude—absorb some of this infrared radiation and help keep us warm. Because of this natural greenhouse effect, the Earth is about 59°F (30°C) warmer than it would be without its atmosphere.

Carbon dioxide, the main gaseous product of the combustion of fossil fuels, is, aside from water vapor, considered the most significant greenhouse gas. It presently constitutes about 1/3,000 of the atmosphere. Other greenhouse gases, such as methane, are twenty times more effective than carbon dioxide at trapping heat, but are present in the atmosphere in much smaller quantities. Plants require carbon dioxide for photosynthesis, and plants and animals release carbon dioxide during respiration. However, the industrial—or anthropogenic—production of carbon dioxide as a product of fossil fuel combustion has significantly increased its presence in the atmosphere. It is estimated that more than 6 billion tons of fossil fuel are burned each year, releasing more than 18 billion tons of carbon dioxide into the atmosphere—the largest amount of waste human activities produce.

Since the Industrial Revolution, the amount of carbon dioxide in the Earth's atmosphere has increased by nearly 30 percent, to over 360 parts per million. (Parts per million is the number of units of one substance present in a million units of another.) If current growth continues, analysts expect the amount of carbon dioxide in the atmosphere to double by 2050. (Another

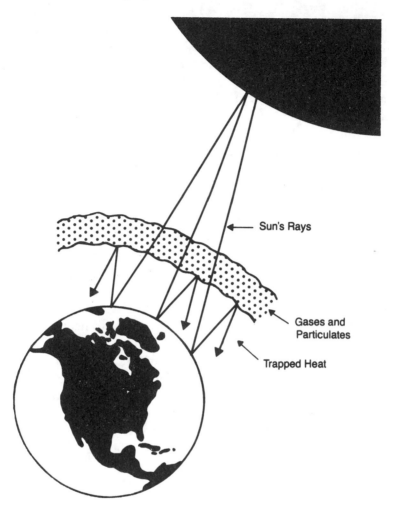

**Figure 9.1** Complex interactions between solar radiation and the atmosphere drive the Earth's climate system. Some of the visible solar radiation that reaches the Earth is absorbed by the land and water, and some is bounced back into space off reflective surfaces, which include snow, ice, and clouds. The Earth also radiates invisible, long-wave heat radiation back into space. Some of this radiation is trapped by greenhouse gases and water vapor, in the same way that glass traps the heat in a greenhouse. It is this effect that has made the planet warm enough for human habitation, and may also be responsible for triggering disruptive climate change.

contributor is worldwide deforestation, which both contributes carbon dioxide when trees are burned and deprives the biosphere of another method of removing carbon dioxide by photosynthesis.) Some scientists even predict that the warming planet will release additional carbon dioxide and methane presently trapped in the ground, creating an even faster warming cycle.

## MAJOR GREENHOUSE GASES AND THEIR SOURCES

Carbon dioxide ($CO_2$)       Fossil fuels, burning forests, animal
                              respiration
Methane ($CH_4$)              Cattle (gas released when food is digested),
                              rice paddies, gas leaks, termites, mining
Nitrous oxide ($N_2O$)        Fossil fuels, deforestation
Ozone and other trace gases   Engine exhausts, power plants, photochemi-
                              cal reactions with other elements, solvents
Chlorofluorocarbons           Air-conditioning, some solvents, some
                              chemicals used in refrigeration

## Is Global Warming Here Already?
## Lessons from the Little Ice Age

Making generalizations about the Earth's climate is a complex task because there have been natural climate changes throughout history and
prehistory. Between A.D. 600 and 1400, for example, European and North
Atlantic climates are believed to have been as much as 1.5°C higher than
today. During this medieval warm period, Viking settlers established
thriving colonies in Iceland, Greenland, and Newfoundland. Agricultural
records, as well as pollen, water level, and fossil data, reveal that a "Little
Ice Age" followed and continued until 1650, causing crop failures,
famines, and the abandonment of northern farms and villages.

Even in this century, some scientists contend that we are entering another ice age and that this climate change is caused by variations in sunspot
activity. However, there appears to be a broad consensus that global warming is upon us. Federal climate researchers say that 1997 was the hottest
year on record, and that nine of the warmest years in history have occurred
since 1986. How much hotter? The estimate is .5°C, the same temperature
difference in the opposite direction that resulted in the Little Ice Age. This
estimate may be low: preliminary figures for 1998 indicate even higher temperatures. Researchers have reported that for the first five months of 1998,
the average global surface temperature was 1.76°F above average for the
benchmark period of 1961 to 1990. Some scientists point to localized "fingerprints" of climate change. One of these is the apparent warming of
Antarctica, where a piece of the ice shelf the size of Rhode Island recently
broke off into the sea. Another is the seeming retreat of Alpine glaciers, exposing ice and rocks that have been buried for thousands of years. Yet an-

other is an increase in unpredictable severe coastal weather, such as large, wet storms in the northeastern United States and severe hurricanes in the Caribbean. The apparent effects of global warming are even being felt in the northern reaches of Alaska. In 1998 an article in the *Wall Street Journal* told of the good fortune of farmers, who in recent years have begun to enjoy the benefits of an extended growing season.

There appears to be general agreement that the Earth's climate is slightly warmer and that carbon dioxide concentrations in the atmosphere are increasing, but that is where consensus ends and intense controversy begins. Disagreement abounds on the rate of warming, on the ability of the Earth to absorb the additional carbon dioxide, and on the effect of global warming upon the Earth's climate.

## Pollution as Coolant

One current opinion is that the rate of warming so far is lower than predicted because of the presence of sulfate aerosols (or fine particles of matter) released by burning high-sulfate coal and erupting volcanoes. The cooling effect of aerosols in the atmosphere was demonstrated in the aftermath of the Mount Pinatubo eruption in 1991, when huge quantities of sulfur gases shot into the atmosphere, where they were converted into sulfuric acid droplets. Only after four years, when the aerosols from Pinatubo were washed out of the atmosphere, did the apparent warming trend return. The presence of sulfate aerosols also explains why temperatures have gone up less than predicted in the more industrialized northern hemisphere than in the southern hemisphere, even though there is more ocean in the latter. Oceans have a larger capacity to absorb heat, and also act as carbon dioxide "sinks," absorbing about 50 percent of the carbon dioxide released into the atmosphere. The aerosols do not mix rapidly between the hemispheres, unlike carbon dioxide and some of the other greenhouse gases, so their effects can be localized.

## It All Depends on Feedbacks . . .

Scientists grappling with the task of predicting the effects of global warming develop General Circulation Models, which are large computer simulations of climatic phenomena. These models combine information about known global weather patterns, such as ocean currents and prevailing winds, with the basic laws of physics to represent the large circulations and interactions of the atmosphere. Scientists then run simulations to try to predict what effect different factors will have. According to the Electric Power Research Institute, the amount of data that must be dealt with in climate modeling is so huge that it takes the world's fastest supercomputers sixteen days to complete a single one-hundred-year simulation. Among the numerous variables they have to contend with are negative and positive feedbacks. For example, carbon dioxide acts as a fertilizer that makes some plants grow faster. As the concentration of carbon dioxide in the atmosphere increases, these plants take more carbon dioxide out of the atmosphere. This negative feedback would slow the rate of warming. On the other hand, as the Earth warms, ice and snow melt, reflecting less light energy from the sun. This would result in a positive feedback and an increase in warming. Scientists also find it hard to predict what effect clouds and water vapor have on global warming. Clouds block about 30 percent of solar radiation coming in, an effect known as *planetary albedo*. However, water vapor is also part of the atmospheric blanket that helps to trap outgoing infrared radiation.

---

### PROS AND CONS

Controversy about the likelihood of global warming centers around its possible deleterious effects. The issue has its share of optimists who even predict beneficial effects for global warming. The following are some of the arguments and counterarguments one is likely to encounter:

Optimists: The increase in carbon dioxide will act as a fertilizer, stimulating the growth of plants. These plants will absorb the excess carbon dioxide.

Pessimists: Plant matter (or biomass) cannot absorb these enormous quantities of carbon dioxide—unless you plant trees over an area of 3 billion acres. In addition, plants and animals produce carbon dioxide in cell respiration. It has also been found that some plants "shut down" their absorbing capacity if they are exposed to high concentrations of carbon dioxide.

Optimists: As the Earth gets warmer, the growing season in high latitudes will be extended, and there will be an increase in the amount of arable land. Global warming will produce an agricultural bonanza. We will be able to grow pineapples in Wisconsin.

Pessimists: Lands in high latitudes, such as northern Canada, do not have the topsoil to sustain intensive cultivation. In addition, the expected speed of global warming will not give natural systems a chance to adapt to change. What we are more likely to see is an environmental mismatch, with unknown consequences.

Optimists: Aerosols will continue to slow the pace of warming.

Pessimists: Aerosols do not remain in the atmosphere as long as carbon dioxide. Carbon dioxide remains for at least one hundred years, while aerosols can wash out in days.

## Waterworld, or an Agricultural Bonanza?

Nobody is really sure what the effects of global warming will be—which, of course, is one of the reasons the issue is so controversial. International climatological organizations predict that by the mid-twenty-first century there will be increases in rainfall, a decrease in sea ice, warmer Arctic and Antarctic zones, and a rise in global sea level. Other local climate changes could include the following:

- The shift of more temperate weather conditions toward the poles.
- The spread of deadly tropical diseases like malaria.
- Different weather in important grain-growing regions, such as the midwestern United States, which could disturb global food supplies.
- Unpredictable weather—probably wetter in coastal regions because of the increased evaporation of water from the ocean, and dryer in the interiors.
- More severe storms.
- Compromised freshwater supplies as rising sea levels encroach further inland.
- A greater reliance on air-conditioning, requiring more electrical energy and therefore, ironically, with current technology, increased reliance on fossil fuels and the resulting increased production of greenhouse gases.

Analysts expect that industrialized countries will be able to adapt to these climate changes, albeit at enormous cost. There is considerably more concern for less developed countries, where agriculture is more traditional and attuned to weather cycles. There are arguments that the consequences of global warming, though serious, are not sufficiently so to force a radical change in our carbon-based energy economy. Others contend that the most urgent reason for trying to reverse or slow down global warming is the very uncertainty of the outcome, which will be difficult to reverse because carbon dioxide and other greenhouse gases have a long atmospheric life span. They point out that we are trying to make predictions about how a complex system will respond to unprecedented change. Moreover, the world's climate is a somewhat mysterious and finely balanced system in which even small changes can have large consequences. There are concerns, for example, that rising oceanic temperatures could shift ocean currents with unpredictable effects. Global warming, say environmental scientists Bernard J. Nebel and Richard T. Wright, has the potential to alter "the basic parameters of the biosphere," affecting not just one ecosystem in a limited area, but "the balances between all the species and ecosystems on earth."

## LONGER THAN A LIFETIME
*Greenhouse gases and their average residence time in the atmosphere*

| Gas | Average concentration 100 years ago | Approx. current average concentration (in parts per billion) | Residence time (in years) |
|---|---|---|---|
| Carbon dioxide | 290,000 | 360,000 | 100 |
| Chlorofluorocarbons and halocarbons | 0 | 3 | 60–100 |
| Methane | 900 | 1700 | 10 |
| Nitrous oxide | 285 | 310 | 170 |

Reversing global warming will require unprecedented international cooperation and resolve that has yet to materialize. So far, a series of international conferences on global warming has failed to produce any binding commitments to cut carbon dioxide emissions from major industrial countries. Advocates seeking a global effort to reduce carbon dioxide emissions point to the international agreement to totally phase out the production of ozone-depleting chemicals (see page 257). However, reducing carbon dioxide emissions is a far more complex and politically charged challenge, because chlorofluorocarbons, the chemicals implicated in ozone depletion, can be replaced with relatively little disruption. Most industrial and commercial interests and fossil-fuel-producing countries are firmly on the side of preserving the status quo. But momentum for change is developing. This includes pressure from the powerful insurance industry, which in recent years has paid out billions of dollars in catastrophic weather-related claims. Insurance industry representatives are reportedly most perturbed by the prospect of unpredictable changes in weather patterns. Also joining the chorus are thirty-six island nations, which foresee obliteration if sea levels rise.

---

### COAL AND CARS

**It's going to be hard to give up on coal . . .**
The world has very large coal reserves. First used by the Romans in A.D. 100, coal is the cheapest reliable source of energy. Sadly, from the standpoint of our atmosphere, it is a detrimental energy source. Trapped methane escapes during the mining and processing of coal, and residues often result in the acidification of local surface water. When it is burned, coal produces carbon dioxide, carbon monoxide, hydrocarbons, nitrous gases, sulfur dioxide, soot, and several trace metals, including mercury.

**And on the internal combustion engine . . .**
Although Henry Ford did his share to popularize car ownership in the 1920s, it was only by the 1960s, with the tremendous proliferation of cars, that pollution from gasoline engines became a problem. In the United States, the transportation sector exceeds power generation as a source of carbon dioxide. On average, a gallon of gasoline contains about five pounds of carbon. Burning gasoline also produces other pollutants, including carbon monoxide and a wide variety of hydrocarbon compounds.

---

## Out of the Black: Toward a Low-Carbon Energy Economy

Faced with the prospect of disruptive climate change, some prominent scientists have proposed large-scale projects—termed *geo-engineering*—to help cut down on global warming and greenhouse gases. Among the proposals are to fertilize the oceans with added iron, which may cause phytoplankton in the top layers of the ocean to absorb more carbon dioxide.

But most environmental and atmospheric scientists are wary of the unintended consequences of grandiose engineering solutions, and believe that the most prudent response to the threat of global warming is prevention: reducing the use of fossil fuels and eventually ceasing their use altogether. Despite the apparent enormous resistance to "decarbonizing" the worldwide energy economy, many observers believe that change is both inevitable and possible. The impetus, they believe, will be a combination of environmental concerns and technological innovations, not resource depletion. Among the features of a low-carbon economy will be high energy efficiency, which will have the added advantage of lowering energy costs; a decentralized system of power generators whose waste heat would be used by homes and industry; and increased reliance on methane gas to replace oil and coal. Among technologies already available are gas turbines, wind generators, rooftop solar cells, and energy-efficient lightbulbs. Other technologies, such as fuel cells and flywheels, soon will be. The ultimate goal would be for the bulk of the world's energy supply to be provided by emission-free sources like solar power and hydrogen gas.

## Tapping the Conservation Reserve

The industrialized world has already quite significantly "decarbonized" its energy diet; it has been estimated that industrialized countries are three times more efficient in their use of energy than the developing world. Appliances and cars are already far more energy-efficient than they were in previous decades. Moreover, many machines—the computer, the washing machine, and the microwave oven, for example—are smaller and consume far less energy than their predecessors did. Nevertheless, energy consumption in the developed world is still increasing.

Some energy experts estimate that we could be using energy even more efficiently, and that at least 80 percent of the "conservation reserve" remains to be tapped. For example, a large proportion of the energy used in the United States goes to heating, cooling, and ventilation, and there could be great gains if buildings were better insulated and designed to use "natural climate control." Houses could, for example, be shaded by the leaves of deciduous trees in the summer, and warmed by the sun when the leaves fall off in the winter. Another area of virtually "instant" energy saving is in lighting, which accounts for 20 to 25 percent of all electricity sold in the United States. Using natural light from windows to illuminate offices in the daytime, or substituting fluorescent lights for incandescents (the former can be up to five times more efficient than conventional lightbulbs), would cut back on our electricity demands for lighting by 75 to 80 percent.

Energy experts also contend that the efficiency of appliances and cars could be significantly improved using available technology. The Worldwatch Institute, an environmental advocacy organization, estimates that if the average mileage obtained by new cars in the United States was raised to 32.5 miles per gallon, U.S. emissions of carbon dioxide would decline by about 250 million tons per year. And of course, huge energy savings could be realized if Americans could be persuaded to fund—and make more use of—energy-efficient mass transportation systems such as railroads, and reverse current trends of extending suburban sprawl.

## Cogeneration

One of the most effective methods of pollution prevention is ensuring that energy in the form of electricity, or heat, is utilized to the fullest extent possible. One method is cogeneration, or using a single energy source to produce both electrical and heat energy. In Linden, New Jersey, for example, a plant operated by Cogen Technologies pumps out 645 megawatts of electricity for New York City, and 1,000,000 pounds of steam per hour to a petroleum refinery. This plant uses combined-cycle gas turbines (which use the exhaust gas from the gas turbogenerator to operate a conventional steam generator). Whereas ordinary power plants are at best 40 percent efficient in their use of energy, the Linden plant is estimated to have an efficiency of over 65 percent.

---

**EFFICIENT ENERGY**

Any energy producer must contend with the two basic laws of thermodynamics: The first, the law of conservation of energy, states that energy is neither created nor destroyed, but may be converted from one form to another. The second Law, also known as the law of entropy, states that in any energy conversion, you end up with less ability to do work. Systems go toward increasing entropy, or disorganization. This process releases heat. The ratio of the amount of energy that is produced in the conversion from one form of energy to another is its efficiency. Using coal to produce electricity, for example, has an efficiency of only 30 percent to 40 percent—nearly 300 calories of coal must be burned to produce 100 calories of electricity. Newly developed gas turbines, which use sophisticated steam cooling systems to cut pollution emissions while permitting high "firing" temperatures to produce the high-pressure gases that drive the turbines, have reportedly attained 60 percent efficiency, which is considered a major breakthrough.

---

## Making the Transition with Methane

Although methane, the main component of natural gas, qualifies as a hydrocarbon, a fossil fuel, and a potent greenhouse gas, supporters promote it as the most suitable fuel to ease the transition from coal and oil. Methane is the product of both the prehistoric and ongoing decay of organic matter in oxygen-free—or anaerobic—conditions. Ancient plant matter buried deep beneath the earth generated methane, as does garbage buried in a landfill. Methane is available in large quantities in many locations worldwide and also, unlike other fossil fuels, can be continually generated and tapped from landfills and especially constructed tanks, called *biodigesters*, where organic matter decays.

Methane has a simple chemical structure (four atoms of hydrogen and one of carbon). When it is burned, it produces carbon dioxide, water, and nitrogen oxides, which makes it significantly less polluting than coal or crude oil. Though still a source of carbon dioxide, methane produces about 50 percent less for the amount of energy generated than coal. There are concerns that methane is a powerful heat-trapping gas, and that if it is handled carelessly and allowed to escape into the atmosphere, its increased use will cause more global warming. However, methane has a far shorter atmospheric life— just ten years, compared to one hundred years for carbon dioxide.

## FROM JET ENGINES TO KILOWATTS: THE RISE OF THE GAS TURBOGENERATOR

Natural gas is being used to generate electricity both in conventional steam-generating plants and in a new generation of high-efficiency gas turbogenerators, which are leading a trend away from giant—and hugely expensive—power plants toward more decentralized energy systems.

A jet engine consists of a rapidly spinning blade that forces air at high pressure into a combustion chamber. There, heated by burning fuel, it creates the thrust that propels an aircraft forward. Brought to earth, mounted on the ground, and attached to a generator, the jet engine becomes a gas turbine. In the past two decades, gas turbines have become increasingly popular worldwide. Combined-cycle generating plants, in which excess heat from the exhaust gases of the gas turbine is used to power a steam turbine, boost efficiency. New technology, which increases the amount of heat energy transferred from the gas to the steam cycle, will improve the efficiency of combined-cycle systems still more. These generators are inexpensive to build, and can be constructed rapidly.

Environmental advocates strongly favor gas turbine plants over conventional oil or coal plants; they emit no sulfur and few particulates. Emissions of nitrous oxide are 90 percent lower, and those of carbon dioxide are 60 percent lower. Energy experts predict even greater improvements in cost and efficiency with advanced metals, new blade designs, and increased gas compression.

A promising area of gas turbine technology is in the use of *biomass*—industrial and agricultural residues and municipal solid waste—as a fossil fuel alternative. Biomass is used to generate electricity with conventional steam turbine power systems. With more efficient energy-generating technologies, experts predict the development of plantations dedicated to energy crops like hemp. Although burning any kind of biomass produces carbon dioxide, proponents of cultivated biomass contend that using biomass will mitigate global warming because growing plants absorb carbon dioxide during photosynthesis. The result, they say, is a "closed" carbon cycle. However, there are concerns that growing plants for energy generation will occur at the expense of agricultural production, which could seriously affect the availability of food supplies worldwide.

## Harvesting the Wind

One pollution-free source of energy is wind power. From the 1100s to the early part of the twentieth century, tens of thousands of windmills ground corn and pumped water on the farms of northern Europe. The first electricity-generating wind turbine was developed in Denmark in the 1890s. This small Scandinavian country has made a major commitment to using wind power to reduce its carbon dioxide emissions. Already, more than four thousand sleek, white turbines resembling large airplane propellers tower over the Danish countryside. New offshore wind farms that are projected to supply more than a quarter of the country's electricity are planned. Currently, wind turbines provide more than 7 percent of the country's electricity, and turbines designed and built in Denmark are now found in "wind farms" around the world.

Wind turbines are relatively simple, and require little maintenance. In a typical generator, turbines with blades 20 to 30 meters in diameter are housed in front of a nacelle, the housing that encloses the motor. The blades are turned to face the wind by a computer-controlled electric motor. Wind turbines use aerodynamic lift—the blades are angled across the wind, and may move at ten times the wind speed—to extract energy from the wind. The blade hub is connected via a low-speed shaft and gearbox to a generator that is synchronized to the same frequency as the electric supply system, or grid. Power control is achieved either by varying the angle, or pitch, of the blade, or by controls to stall the blades if the wind speed exceeds the rated power of the generator. Machines now in development have lighter and more aerodynamic blades, improved rotor-hub connections and drive trains, new types of electronic blade controls, and more advanced power electronics, including some that allow the turbines to operate within a greater range of wind speeds.

In past decades, wind power was far more expensive than traditional fossil fuel sources, but as the reliability and efficiency of wind turbines has improved, the cost of wind-generated electricity has come down to similar levels for power plants fueled by natural gas or coal. Moreover, wind power is free of the environmental price tag carried by traditional generating plants—a price that is generally not factored into the cost. And while wind power currently provides less than 0.1 percent of the world's electricity, the number of wind turbines is increasing yearly, and wind-power enthusiasts predict a vastly more significant future. By 1995 the combined output from European wind turbines generated enough power for 1 million homes, cutting over 4 million tons of carbon dioxide emissions. The European Wind Energy Association estimates that 10 percent of the European Union's electricity could be generated by wind power by the year 2030.

Wind turbines can be noisy. The sight of thousands of wind turbines on a ridgetop can be unappealing, and the blades can kill birds. In addition, there are unsolved logistical problems. In some cases, long distances separate the world's large wind resources from the major population and industrial centers where most energy is used. In the United States, nearly 90 percent of the country's huge wind potential is in the Great Plains, hundreds of miles from Chicago and about 1,500 miles from New York. In other cases, geographic location is an advantage. The Worldwatch Institute has estimated that the La Ventosa (literally, "windy") region near Mexico City has enough wind potential to provide one-third of the country's electricity. At least twenty small subtropical island countries have nearly constant trade winds that could meet a large share of their electricity needs at a substantial saving over expensive diesel generators.

## From Dreams to Reality: Solar Energy Goes Mainstream

In the search for an inexhaustible, nonpolluting source of energy, thoughts always turn to harvesting the sun. It is estimated that if just 0.01 percent of the solar energy reaching the whole surface of the Earth could be harnessed to perform useful work, it could supply all human needs with no deleterious environmental effects.

Every source of energy exploited by humanity so far has been highly concentrated. But sunlight is a diffuse source—it falls evenly over a wide area. The challenge is to develop economical and efficient methods of concentrating solar energy into a form, such as fuel or electricity, that we need for heat and to power vehicles, appliances, and other machinery. In fact, the technology to capture the sun's energy in various ways is advancing rapidly. The present challenge is to develop manufacturing and generating methods that would make solar energy economical.

One solar technology that is already well established is solar water heating. A flat plate solar collector for heating water consists of a thin board "box" with a glass or clear plastic top and a black bottom with water tubes embedded in it. Faced toward the sun, the black bottom gets hot as it absorbs sunlight. Water circulating through the tubes is heated and goes to the storage tank. The heated water may be moved by a pump, or by natural convection currents. For the passive solar water-heating system, the system is mounted with the collector lower than the tank: heated water from the collector rises into the tank while cooler water from the tank descends into the collector.

## THE MOTHER OF INVENTION

Despite being just a short distance from the world's largest oil reserves, Israel has no significant fossil fuel resources of its own. Cut off from its Arab neighbors, and with limited financial resources, the country made an early commitment to solar energy. Israel has become a leader in solar technologies, establishing its first solar company in 1953. The tiny Middle Eastern nation has the world's greatest density of solar collectors and one of the highest concentrations of solar scientists and engineers. On Israeli rooftops, nearly a million solar collectors provide hot water for more than 80 percent of Israel's homes.

### Cooking with the Sun

For hundreds of millions of people in the developing world, cooking is often the largest use of energy, and widespread deforestation has reduced the availability of firewood. Using glass, aluminum foil, glue, and cardboard, researchers have developed a solar cooker that consists of an insulated box with a transparent, removable top. By adding a reflector to increase the sunlight entering the box, interior temperatures can top 100°C, enough heat to cook the three to four pots that fit inside. If the sun is strong, rice can be cooked in two hours, corn in three to four, and beans in five to six. There have been problems overcoming cultural preferences, such as for food cooked quickly on an open fire, but solar boxes are catching on in areas where there is little firewood available.

## Electricity from Solar Generating Sources

The key to more extensive reliance on solar energy is developing a cost-effective means of converting sunlight into mechanical energy, electricity, or chemical fuels. To date, two methods have emerged as economically viable: photovoltaic cells and solar trough collectors.

## Photovoltaic Cells

Solar photovoltaic cells are semiconductor devices made of silicon that convert the energy from sunlight into moving electrons. Each solar cell consists of two very thin layers of material. The first layer is made from a material called negative-type silicon, which is specially treated so that it has more electrons than normal silicon. The second layer is made from positive-type silicon, which has gaps in its structure because it has fewer electrons. The kinetic energy of light striking this "sandwich" dislodges electrons from the first layer, creating the potential for an electric current to move between the two layers. The solar device is hooked up so that electrons from the lower side flow through a motor or other electric device back to the upper side. Photovoltaic cells do not store electricity; independent solar systems that provide electricity when the sun is not shining are equipped with a storage device like a lead-acid battery.

Advances in the technology are leading to higher efficiencies and lower costs. Early photovoltaic cells used crystalline silicon more than 100 microns thick. New thin-film technologies utilize silicon with a different non-crystalline structure (called amorphous silicon) or other semiconducting materials like cadmium telluride and copper indium diselinide. Researchers are also experimenting with combinations of amorphous and microcrystalline silicon. To increase efficiency by using as much light energy as possible, scientists are also using combinations of materials that release electrons in response to different wavelengths of light. For example, gallium arsenide responds to blue light, while silicon reacts to the red portion of the spectrum. These methods create layers of active photovoltaic material just one or two microns thick that can be economically deposited onto a large piece of material like glass. These technologies enable large-scale production, a factor that has brought down the cost of photovoltaics by a factor of fifty in the past twenty years. Producing solar cells resembles the manufacturing methods for making silicon computer chips, and solar proponents predict that the technology will benefit from the economies of large-scale fabrication developed by the semiconductor industry.

It is expected that flat-plate photovoltaic technology will have its largest potential market in "distributed grid" applications where the installations are near the user, rather than in a central power station that uses long transmission lines. Photovoltaics are suitable for distributed power because there are no economies of scale (the size of the plant does

## FROM OUTER SPACE TO OFF-THE-GRID

The photoelectric effect—that light falling on certain materials could cause a flow of electrons—was first discovered by the French scientist Alexandre-Edmond Becquerel in 1839. Within fifty years, scientists had began to manufacture primitive solar cells out of selenium, but high cost and low efficiency made these devices useful only for photographic light meters. In 1954, scientists at Bell Laboratories discovered that silicon would produce electricity when exposed to sunlight. The initial solar cells were enormously expensive, but the research and production got a boost when U.S. scientists turned to the solar cell to power satellites. By 1980, photovoltaic modules were cheap and efficient enough for use in remote applications where small amounts of power were needed. In the mid-eighties, Japanese electronics companies gave solar technology a boost by attaching tiny solar cells to devices like handheld pocket calculators.

One of the most rapidly expanding markets for photovoltaics is in rural areas. The cost of photovoltaic cells has dropped to the point that it is sometimes now more economical to install an array of photovoltaic cells in a village in a developing country than to build a centralized power plant plus the grid (the network of power lines taking power from generating stations to customers) needed to deliver the power. In the past few years, hundreds of thousands of homes in Africa, Asia, and Latin America have been electrified using solar cells. Solar cells can also be used to charge telephone batteries, making it possible to introduce wireless communications to people living in areas where conventional telephone service is unavailable.

not affect the cost of electricity generated), operators are not needed on site, and there is no noise or pollution.

## Power Towers and Solar Troughs

Another approach to tapping the sun's energy is to concentrate sunlight onto a tank holding a liquid that is heated to create steam and drive a generator. The first large-scale application of this technology was Solar I, a 77-meter high, tower-mounted receiver built in southern California's Mojave Desert in the early eighties. The tower is surrounded by seven hectares of mirrors (called heliostats) that reflect sunlight onto the top of the tower, where water is quickly vaporized into steam used to power a 10-megawatt generator on the ground. The facility turned out to be expensive and unreliable, and was taken out of service. But the engineers who designed it are working on similar projects.

Another method of concentrating the sun's rays is to use a parabolic solar collector that is coated with a mirrored surface to reflect light coming from different angles onto a single point or tube. These solar trough systems have benefited from such inventions as inexpensive reflective materials, improved heat-transfer fluids, more efficient solar receivers, and electronic tracking devices. One system has mirrored troughs nine feet high and forty feet long that concentrate the sun's rays onto oil-filled tubes that run parallel to the mirrors. The troughs are mounted on a device that allows them to follow the arc of the sun, keeping it focused on the collection tube. To generate electricity, the fluid heated by the solar trough is circulated to a power station. There it heats water to produce steam, which then powers an electricity-producing turbine.

## Tapping into Topography: Hydroelectric, Geothermal, and Tidal Power

Other emission-free methods of power generation take advantage of natural features of the Earth. Before the advent of steam engines heralded the Industrial Revolution, water wheels were used to power a range of enterprises from grinding corn to powering a weaving loom. Hydroelectric power, which also harnesses the kinetic energy of falling water, provides one-fifth of the world's electricity as well as being a source of primary—or direct— energy in water wheels. To produce hydroelectric power, water under high pressure—at the base of a dam or at the bottom of a pipe from the top of a waterfall—is used to drive a hydroturbogenerator. This traditional power source is dependent on topography and huge amounts of water. Norway, land of the scenic fjord, gets 95 percent of its power from falling water. The United States and Canada generate 13 percent of the world total. Although relatively emission-free, large hydropower plants exact an enormous ecological, social, and cultural toll. Thousands of people are displaced when huge dams are created to power hydroelectric projects; often, as in the development of Egypt's Aswan High Dam, important archeological artifacts are also destroyed. Hydro-Quebec's grandiose James Bay project, which taps the hydropower potential of three major river systems in the vast wilderness of northwestern Quebec, has threatened the traditional hunting and gathering culture of the native Cree and Inuit peoples and compromised a pristine wilderness. Worldwide, plans for large hydropower developments are coming up against local and international resistance. Moreover, there are concerns that large hydroelectric projects can contribute to global warming because they are a significant source of methane generated as the huge amounts of vegetation submerged by the creation of a dam decay.

Two other methods of power generation that exploit natural features of the Earth are geothermal and tidal power. In several locations, the molten rock of the Earth's interior is close enough to the surface to heat groundwater. Geothermal facilities use such naturally heated water or steam to heat buildings or drive turbogenerators in such countries as Nicaragua, the Philippines, Kenya, and Iceland. However, geothermal power may not be sustainable. For example, output from The Geysers, a geothermal facility north of San Francisco, has dropped because tapping the hot steam is depleting the groundwater. There are also potential pollution problems, because the water and steam are frequently laced with salts and other contaminants, particularly sulfur compounds, leached from minerals in the bedrock.

Tidal power tries to capture the pollution-free energy in the twice-daily rise and fall of the tides. A tidal power plant would consist of a dam built across the mouth of a bay, with turbines mounted in the structure. The incoming and outgoing tides generate power as they flow through the turbines. This method of generating power requires very high tides, and there are only about fifteen locations in the world where the tides are high enough for this kind of use. Large tidal power plants already exist at two of these places, in France and Russia. There are also concerns about adverse environmental impacts, including altering the circulation of saltwater and fresh water in estuaries.

## Carbonless, But Costly: The Nuclear Alternative

Faced with the twin specters of global warming and resource depletion, at least four countries—France, Belgium, South Korea, and Japan—have made major commitments to nuclear power. In 1995, France produced 76 percent of its electricity from nuclear power, and Japan plans to increase its nuclear percentage from 24 percent to 43 percent by 2010. Both Belgium and South Korea rely on nuclear power for more than half of their energy needs. Proponents contend that taking the nuclear route does not contribute to global warming and actually would protect the environment, particularly in fossil-fuel-poor areas where it is likely that the local population will chop down trees for fuel. Moreover, they note, there is enough uranium to fuel nuclear reactors well into the twenty-first century, with the possibility of extending the nuclear fuel supply by reprocessing.

Currently the United States gets about 20 percent of its power from nuclear reactors. But, in part because of public concerns about safety that have resulted in heavy political pressure, the U.S. nuclear power industry is in precipitous decline. The last order for a nuclear power plant that was not canceled was placed in 1974. In 1989 the $5.5 billion Shoreham nuclear power plant on Long Island, New York, was turned over to the state of New York for dismantling without ever having generated a single watt of power. Throughout the country, the number of operating plants peaked at 111 and is now dropping, with at least 12 plants slated for early shutdown as utilities contemplate huge costs to fix problems. In less developed countries, however, the demand for energy is growing, and these countries may turn to nuclear power. In fact, currently, the only continent placing new orders for nuclear plants is Asia.

At present, all nuclear power plants use fission to generate electricity. Fission is a nuclear reaction in which a heavy element, such as uranium, is split into smaller elements after being struck with a free neutron. Uranium "splinters" readily, releasing two or more neutrons, which in turn strike and splinter other uranium nuclei in a chain reaction. The uranium is located in the reactor vessel, the central core of a nuclear power plant. The reactor is designed to support a continuous, controlled, chain reaction that is achieved by surrounding the enriched uranium with a moderator (or coolant) and using control rods to slow or stop the reaction. The control rods, usually boron or cadmium, absorb the free neutrons that would cause the nuclear chain reaction.

In nuclear plants in the United States, the moderator is very pure water, and the reactors are called light water reactors. Most of these plants use a double loop: the nuclear reaction does not directly produce electricity. The primary coolant water (or moderator) is superheated by circulating it through the reactor, but it does not boil, because the system is under very high pressure. This superheated, radioactive water is circulated through a heat exchanger, where it boils other water, the secondary coolant, to produce the steam used to drive the turbogenerator. The primary coolant, because it circulates in a sealed system, does not mix with or contaminate the secondary coolant.

## Weighing the Environmental Load

Compared to coal- or oil-fired power plants, the nuclear option has some striking environmental advantages. Compare, for example, a 1,000-megawatt nuclear plant with a coal-fired plant of the same capacity, each operating for one year: The coal plant will consume about 3 million tons of coal, emit over 10 million tons of carbon dioxide, and produce about 100,000 tons of ash, requiring land disposal. The nuclear plant requires about 30 tons of enriched uranium, obtained from mining 75,000 tons of ore, emits no carbon dioxide or acid-forming pollutants, and produces about 250 tons of radioactive wastes. In the event of a catastrophic accident, however, the otherwise more environmentally benign nuclear plant could cause widespread radiation sickness, scores of human deaths, cancers in humans and animals, and widespread, long-lasting environmental contamination. And the 250 tons of radioactive waste require long-term safe storage and disposal.

## WHAT HAPPENED AT CHERNOBYL?

On April 26, 1986, the catastrophic explosion of a nuclear reactor at Chernobyl, in the former USSR, killed thirty-three people directly and contaminated a broad area of the republics of Belarus, Russia, and Ukraine. The accident was the result of gross operator error compounded by a lack of safety features in the plant. (The Chernobyl reactor, for example, had no containment building.) At 1:24 A.M. local time, electrical engineers who were unfamiliar with nuclear safety systems tried to test the reactor at low power. When they realized that this low power level could lead to trouble, they tried to get the power back up by disabling major elements of the safety systems. Instead of simply shutting down the reactor, they withdrew almost all of the control rods that would have slowed down the reaction. Power increased so rapidly that within seconds the reactor went from 6 percent power to thirty times normal full power. A meltdown and a fire occurred, a steam explosion blew the top off the reactor, and huge quantities of radioactive materials spewed thousands of feet into the atmosphere. Two of the engineers were killed in the explosion; thirty-one workers brought in to seal off the reactor after the explosion died of radiation sickness within six months. Already, more than five hundred cases of childhood thyroid cancer have been ascribed to Chernobyl. It is estimated that the accident and its aftermath exposed 5 million people to various levels of radiation. In the Gomel region of Belarus, an area of nearly 2 million people, the toll in uprooted lives and a pervasive fear of the future has been enormous.

Thus, the main concerns about nuclear power are the disposal of radioactive wastes and the potential for catastrophic accidents. The biggest problem in a nuclear plant is controlling the amount of heat in the radioactive core, and the most feared accident is a meltdown, in which coolant is lost and the core is overheated. Some proponents of nuclear power claim, however, that we have the technology to build inherently safe nuclear reactors designed to automatically quench the chain reaction and suppress the heat of nuclear decay. Engineers are also designing simpler, smaller power plants. One design for so-called advanced light water reactors has safety systems that use natural forces like gravity and convection that are impossible for operators to inactivate. Another design, by South Africa's electric utility, Eskom, is for a "pocket-sized" nuclear reactor that does not use water. The fuel for this small reactor consists of particles of uranium embedded in carbon balls. The fuel heats up helium, an inert gas, which both cools the reactor and drives the turbine. According to the designers, these plants would produce no pollution and store their own waste.

## In the Nuclear Future

Nuclear proponents see even vaster power potential in fusion and breeder reactors. In nuclear fusion, two atoms of a light element, such as hydrogen, are compressed and joined at extraordinarily high temperatures to form a single heavier element, such as helium. High-speed neutrons are ejected in the process and energy is released (the amount is estimated to be one million times more than in a conventional chemical reaction). Fusion makes the sun and the stars burn, and powers thermonuclear explosives like the hydrogen bomb. Proponents contend that controlled nuclear fusion could produce the equivalent energy of three hundred gallons of gasoline from a gallon of sea water without the problem of nuclear waste storage, because the products of a fusion reaction are not radioactive. Other scientists question this, predicting that the products will be radioactive. Moreover, between theory and energy generation lie enormous technical and scientific challenges. Scientists working to adapt the fusion process to nuclear energy production face the problem of achieving the high heat and uniform compression required to initiate fusion. Although some machines have achieved fusion, the energy required to bring about the reaction has so far always far exceeded the energy produced. There are also concerns that the structure housing the reactor would become brittle and radioactive.

Yet another method of generating nuclear power that has some proponents is the use of nuclear breeder reactors. This kind of reactor would generate electrical energy at the same time as it produced fuel for fission reactors. A breeder reactor extracts up to 80 percent of the fuel's total energy, compared to the 1 percent efficiency of a conventional fission reactor. The breeder uses the fission process to convert uranium 238 into plutonium. Concerns about safety and the danger of weapons proliferation led the United States to abandon commercial breeder reactor projects using plutonium in 1970. The Japanese were more enthusiastic proponents of breeder reactors, but in early 1996, a serious accident at the Monju fast-breeder nuclear plant in northern Japan caused its indefinite shutdown. Some analysts, however, expect this resource-poor island nation, which is anxious to achieve energy self-sufficiency, to press on with the program.

## The Quandary of Nuclear Waste

Worldwide, the nuclear power industry is searching for off-site storage for spent nuclear fuel and other radioactive waste. This description of nuclear fuel is actually a misnomer: spent fuel rods are taken out of operation because they have become too radioactive to control. The fission of uranium, or any other heavy element, forms lighter elements, including iodine, cesium, strontium, and cobalt, that are generally unstable isotopes (a form of the element in which the atoms have more or less than the usual number of neutrons). These radioisotopes become stable by spontaneously ejecting subatomic particles or high-energy radiation or both. Any materials in and around the reactor may also be converted to unstable isotopes and become radioactive by absorbing neutrons from the fission process. These products of fission are the radioactive wastes of nuclear power.

When unstable isotopes have finished ejecting particles and radiation, they cease to be radioactive. This process is known as radioactive decay, and the amount of time that it takes half of a radioactive isotope to decay is called its half-life. The half-lives of various isotopes range from a fraction of a second to many thousands of years in the case of plutonium. As long as radioactive materials are kept isolated from humans and other organisms, they decay harmlessly. The challenge of disposing of radioactive materials lies in devising a safe method of storing those wastes until they lose their radioactivity. In the case of some isotopes, such as strontium 89 and iodine 131, the period of time is a few months; the bioactive isotopes strontium 90 and cesium 137, however, which have a half-life of thirty years, pose potentially serious problems.

For short-term containment, the spent fuel is first stored in deep water tanks on the site of nuclear power plants, but many plants are running out of both space and time. The United States and most other countries using nuclear power have decided to bury nuclear wastes underground, but by 1998, no nation had as yet carried out the burial. Many nuclear nations have not even been able to find a site that may be suitable for receiving the wastes, and everywhere, storage proposals face strong local opposition.

## Wanted: A Nuclear Waste Site

Nuclear scientists seeking a place to bury nuclear waste, potentially for tens of thousands of years, look for geologically stable areas that are reasonably remote from human settlement. The burial site must be located among rocks that are impermeable to moisture and far from a source of water, so that there is no potential for the nuclear waste to get into a water supply. The time criterion has been set at ten thousand years. This has proved problematic: everywhere scientists look, there is evidence of volcanic or earthquake activity or groundwater leaching within this time period. And then there is the problem of local opposition. In 1987 the U.S. Congress selected Yucca Mountain in southwestern Nevada to become the nation's nuclear dump. The U.S. Department of Energy, looking to bury medium-level government waste containing long-lasting radioisotopes, has constructed a $1.5 billion Waste Isolation Pilot Plant in salt caves 2,150 feet beneath the desert in southwestern New Mexico. In both cases, the states sued to block the sites. This opposition has led some observers to conclude that radioactive waste disposal is less a function of technical feasibility than a question of social or political acceptability. They contend that the ten-thousand-year criterion is unrealistic, making it virtually impossible to find an acceptable nuclear waste site.

Because of safety concerns and the problems of waste storage, most environmentally oriented energy experts do not support the nuclear alternative. However, nuclear energy may yet become more widely acceptable if the consensus develops that the continued use of fossil fuels is so damaging to the atmosphere and the Earth's climate that limits must be placed on their use.

## Toward the Zero-Emission Vehicle: The Coming Transportation Revolution

In about 1975, automobiles surpassed power plants as the major source of atmospheric carbon dioxide. In addition, "mobile sources," that is cars, trucks, and buses, release the majority of the pollutants that foul our air, including volatile organic compounds (VOCs), carbon monoxide, and nitrogen oxides that lead to ground-level ozone (see chapter 10).

The internal combustion engine, so called because the fuel is burned inside a cylinder, has driven the transportation industry for more than one hundred years. First developed and patented by Nicolaus August Otto in 1866, its superior power-to-weight ratio (compared with a bulky steam engine) facilitated the development of small, efficient engines. Although electric and steam-powered cars posed stiff competition at the turn of the century, gasoline proved the winner. Usurping the internal combustion engine with an emission-free alternative will require massive changes to overcome the inertia of "business as usual" in both the motor vehicle industry and the public. The impetus for such change will have to come from governments: observers believe that only in response to legislation would the international motor vehicle industry consider abandoning such an entrenched and successful technology. In the United States, the first impetus came from the California Air Resources Board, which decreed in 1990 that within eight years, 2 percent of all vehicles sold in the state must be "emission free." However, by 1996, the state substantially watered down this requirement in response to contentions by the auto and oil industries that alternative technologies were not yet commercially viable. But the board has stated it intends maintaining its goal of having zero-emission vehicles account for 10 percent of vehicle sales by 2003. And the automobile industry appears to be moving toward some recognition of the need to phase out the internal combustion engine. In 1998, the Ford Motor Company, in an alliance with Daimler-Benz and Ballard Power, a Canadian manufacturer of alternative power systems, announced plans to produce 100,000 fuel cell–powered cars by 2004. Meanwhile, a limited but growing number of vehicles powered by so-called clean compressed natural gas (CNG), which is mostly methane, have started to hit the roads.

The task of building an affordable, lightweight, high-mileage car that does not pollute the atmosphere is a formidable technical challenge that some experts say will be harder than putting men on the moon. One approach to devising a zero-emission vehicle (ZEV) translates into developing an electric car powered by onboard rechargeable batteries that can provide sufficient power and range to make their operation economical. Electric cars provide an on-demand power system that is quiet and nonpolluting. When stopped, they do not consume power or produce pollutants, unlike conventional cars whose engines idle when they stop at a traffic light or in a traffic jam. Currently electric cars have two main drawbacks: they are expensive and limited in range by present battery technologies. In 1998 the average price tag for a two-seat electric car with a top range of just sixty or seventy miles was $35,000. Commentators noted, however, that the hefty price tag bought a vibration-free ride and a far simpler engine with few maintenance requirements. There are no air filters, spark plugs, or fan belts to change. In addition, these cars are winners on the pollution front. Even counting the emissions from the power plants used to charge the batteries, the average electric vehicle produces one-third of the pollution California will allow under its ultra-low-emission standards.

Factors that will make electric or electric-hybrid cars more affordable and commercially successful include advances in electrical engineering, materials science, aerodynamic design, and low-resistance tires. Among new developments are the use of regenerative braking systems, in which the energy of the car's momentum is used to run the motor in reverse, to generate electricity that is stored in a battery for later use. Another advance will be the use of lightweight composite materials that will come pre-dyed, thus avoiding the need for painting, one of the most polluting aspects of automobile production. At present, however, composite materials are too expensive to be practical, and so car manufacturers are developing other methods of decreasing the weight of cars, including the use of aluminum and high-strength steel.

## Beyond the Lead-Acid Battery

The key to advancing electric cars from environmental correctness to commercial success is either an alternative power source or a breakthrough in battery technology that will produce a readily rechargeable, low-cost, lightweight battery that will store large amounts of power. Such batteries will also have to meet power requirements for acceleration and maintaining high speeds. The conventional lead-acid battery has the disadvantage of high cost, a short lifetime, weight, bulkiness, and limited storage capacity. It has been estimated that 500 kg of lead acid batteries produce the energy equivalent of two to three gallons of gasoline! In search of increased energy storage capacity, researchers are adapting the more efficient but costlier nickel-cadmium and lithium batteries used in computers and other electronic devices. Other researchers are working on a range of new battery materials, like nickel-metal hydride, sodium-sulfur and sodium-nickel chloride. Rechargeables using lithium ions appear to offer much promise, producing four times the energy of an equivalent-sized lead-acid battery. Lithium-ion batteries are also less environmentally damaging, have a longer life, are safer to make and operate, and use cheaper raw materials. Researchers are experimenting with various methods of constructing a lithium battery pack, including using synthetic carbon impregnated with lithium ions. The carbon serves as the negative electrode (anode) from which the electrons are generated in a battery. Another technique creates battery cells made in multiple layers. Each cell consists of five layers that include a current-collecting metal foil, a positive electrode (cathode), an electrolyte, a lithium foil anode, and an insulator. These cells will be connected to form a module, with several modules connected to form an electric vehicle battery pack. This pack is expected to produce an energy-to-weight ratio more than five times that of lead-acid batteries, and to allow electric vehicles to achieve a range in excess of three hundred miles per charge.

Intensive testing and the investment of hundreds of millions of dollars is starting to bring new battery technologies to commercial fruition. The Nissan auto company was expecting to market an electric vehicle powered by lithium batteries in the United States by the end of 1998. Nickel-metal hydride batteries are already in use in a number of production vehicles, including Toyota's Prius hybrid, which uses a combination of gas and electricity. One problem, though, is economics. According to some estimates, it costs an estimated $30,000 to outfit a car with nickel-metal hydride batteries. Another is weight: the battery systems powering the present generation of electric vehicles can weigh over two thousand pounds.

## Using Flywheels for Energy Storage

Another energy storage device that holds great potential—pending the solution of various technical and financial challenges—is the flywheel, a mechanical battery that stores energy in the form of a spinning disk. Some automotive engineers believe that a flywheel system has the potential to store far more energy than any known battery. A flywheel makes very little noise and emits no pollution. No hazardous materials are used in its construction. Operating on the same principle as a potter's wheel, a flywheel disc is set spinning at high speed by an integrated electric motor/generator. It is contained inside a vacuum, eliminating resistance. The ensuing long spin stores kinetic energy, which can then be converted to electricity by a generator. Unlike batteries that can take hours to recharge, flywheels can be revved back up to speed in ten to twenty minutes. Although it was invented over a century ago, the flywheel only became practical with the development of strong, lightweight composite materials in the 1970s and 1980s. Modern composites can spin in a vacuum at up to 200,000 rpm, with the potential to store and release energy at an efficiency of more than 90 percent. However, as with other new technologies, there are expensive hurdles that must be overcome before spinning disks become an economical alternative to gas pumps. One problem is devising affordable ways of suspending the flywheel in a vacuum so that there is no friction. The flywheel also needs to be securely contained, so that in the case of an accident it does not become a lethal flying saucer.

Undeterred by the practical challenges, two California brothers, Benjamin and Harold Rosen, developed a novel hybrid vehicle that is powered by a flywheel combined with a gas turbine engine. The turbine propels the car while it is cruising, while the flywheel provides bursts of power for acceleration. More than five years and $24 million in development, a converted sports sedan was road-tested in January 1997. However, unable to attract major funding from automobile companies, Rosen Motors closed in November 1997.

Although zero emissions are the goal, the more practical and cost-effective alternative of the hybrid vehicle could be more widely acceptable because they have a driving range comparable to gasoline-powered vehicles.

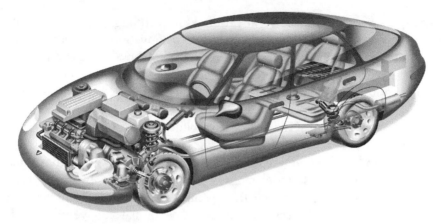

**Figure 9.2** A hybrid electric vehicle can combine a gasoline engine with an electric motor powered by batteries. Switching between gasoline and electricity, depending upon the driving conditions, cuts down on emissions and greatly increases the fuel efficiency of the vehicle. For instance, a battery-powered car is far more fuel efficient in stop-start traffic, while a gasoline engine performs well in high-speed highway driving.

Hybrid vehicles are powered by different forms of fuel, depending on the driving conditions. This makes the vehicles more efficient. In today's cars, only one-fifth of the energy in the fuel is turned into useful mechanical power. The factor that wastes the most energy is operating a car's powerful engine at low power almost all of the time. For most motorists, this is evident when they compare their car's fuel consumption in stop-start traffic to continuous high-speed driving on a highway. One kind of hybrid vehicle uses a small storage device that kicks in when the engine is operating at low power. When the car accelerates to high speed, the gasoline engine switches in. Many hybrids use regenerative power devices to capture and store braking energy in a battery. A hybrid electric car, optimized to operate within a single range of power, can achieve up to 35 percent efficiency. Using a gaseous fuel, hybrids would emit 85 percent less carbon dioxide than today's automobiles, and 95 percent less carbon monoxide and nitrogen oxides. Proponents contend that a well-designed hybrid would have lower emissions than an equivalent battery-powered car that is charged by the kinds of power plants that are in use today. Toyota claims that the Prius, which combines a highly efficient gasoline engine and an electric drive, can travel 70 miles per gallon.

## More Mileage with Fuel Cells

Fuel cells—a highly efficient method of electricity generation that first made its debut in the late 1830s but languished until it got a boost from the

U.S. space program—hold great promise for both large- and small-scale power generation. Fuel cells can be as large as a conventional power plant or small enough to fit into a space shuttle, or under the hood of a car. Although most fuel cells are still assembled by hand and are prohibitively expensive, once economies of scale are achieved, they may become the energy technology that can provide the same kind of performance as the internal combustion engine—with the advantage of zero pollution and noise.

Fuel cells generate power by using a catalyst to create a chemical reaction between hydrogen and oxygen that releases electrons and forms water. As the first—and simplest—element in the periodic table, hydrogen, consisting of one proton and one electron, promises to be an important factor in many developing energy technologies. A fuel cell is like a conventional battery, in that a chemical reaction causes electrons released from one electrode to flow to a second electrode. In a battery, however, the active ingredients are part of the electrodes and are chemically altered and depleted during the reaction. In fuel cells, a gas or liquid fuel such as natural gas, methanol, hydrogen, and even gasoline, is supplied continuously to one electrode, and oxygen or air to the other from an external source. A fuel cell–powered car has a far longer range than an electrical vehicle powered by batteries. The latter can only store a small electrical charge, while the amount of electricity generated by a fuel cell is simply dependent on the size of its fuel tanks.

Because they can achieve power levels comparable to today's internal combustion engines, fuel cell assemblies that are electrically connected in series are considered the most promising candidates for use in electric cars and other applications. These are called *polymer electrolyte membrane cells* or PEM cells. The membrane works by effectively separating the hydrogen from its electron; the proton goes through the membrane while the electron stays behind, thus creating an electric current. This approach to fuel cells has benefited from advances in membrane technology. One method uses a 20-micron-thick membrane made of Gore-Tex, the same material used by outdoor enthusiasts for waterproof windbreakers. In 1998, in a test of PEM membrane technology, a house in upstate New York was taken out of the grid and became the first fuel cell–powered house in the United States. Proponents of alternative energy see the test as a significant event. One government official compared the launching of a fuel cell–powered house to the introduction of the electric refrigerator and the room air conditioner.

## Hydrogen, the Fuel of the Future

Visionaries of pollution-free power have a large place in their plans for hydrogen. This simple element is the basis of fuel cell technology. In addition, conventional cars can be modified to run on hydrogen gas as a fuel in the same manner as they are now beginning to be run on methane. Burning hydrogen produces no hydrocarbon pollution or carbon dioxide; the only by-products are water vapor and nitrogen oxides. In its isolated state, hydrogen is highly flammable, and is only found in compounds such as water. Gasoline is equally flammable, however, and methods of handling it safely have been developed. Petroleum refineries already produce hydrogen from natural gas, where it is used to increase the hydrogen/carbon ratio of petroleum products and to extract sulfur from them. It is also produced in large quantities to make ammonia.

But if hydrogen is to become the fuel that can meet the gargantuan energy needs of the United States, methods will have to be devised for extracting it in far larger quantities. Energy visionaries propose using solar- and wind-derived electricity to separate water into hydrogen and oxygen in giant facilities in the desert areas of the United States, and then using the already installed gas pipeline infrastructure to transport it. They will also have to find convenient and affordable storage methods, especially for transportation. As a liquid, hydrogen would have to be refrigerated to −252°C. If it is only compressed, rather than being stored at the temperature of deep space, a high-pressure cylinder would require thirty times the capacity of the current gas tank.

## Leapfrogging into the Future

By 2050 the earth's population is expected to be at least twice its present size. Most of this population growth will be in developing countries that are already increasing their energy consumption as they industrialize, and it is expected that worldwide energy requirements will be astronomical. Satisfying the world's energy needs while minimizing global carbon dioxide output and environmental damage may be the most formidable technological challenge of the twenty-first century.

There are arguments that the worldwide energy economy is inextricably locked into hydrocarbons and that the consequences for the Earth's atmosphere and climate will not be so severe that drastic changes to the world's energy supply will be necessary. This contention goes with the belief that humankind has the ingenuity to employ technology and science to adapt to potential climate change.

Environmentalists urging for binding international commitments to reverse global warming argue that weaning the world from a carbon energy diet need not be onerous because many of the technologies already exist or are about to become practicable. The solutions, they contend, are more political and social than technical: Worldwide, energy generation and consumption is hugely wasteful. Nigeria, for example, "flares off" as waste most of the natural gas produced as part of its gasoline production. In the United States, probably the most profligate energy consumer on Earth, most Americans appear to consider it their constitutional right to wear a T-shirt indoors in the dead of winter and a jacket indoors at the height of summer. Legislators who favor raising gasoline taxes or requiring American auto manufacturers to meet higher fuel economies do so at the risk of their political careers. Mass transit, far more energy-efficient than the private car, limps along with funding deficits. Analysts assessing the feasibility of alternative energy sources require that those sources meet the same cost criteria as traditional fuels without taking into account the environmental costs of fossil fuel energy.

Nevertheless, the world's economy appears to be evolving toward more decentralized and efficient energy delivery. New power plant orders are likely to be for smaller, highly efficient gas turbines over enormously expensive mega-power projects. Alternative technologies such as wind and solar power appear to have gained a significant place on the energy scene. There are even hopes that developing countries can take advantage of these changes to leapfrog traditional methods of energy generation, avoid the high costs and environmental damage, and meet their energy needs at the same time.

If nothing else, the impetus to stop burning petrochemicals for energy might develop because these hydrocarbons will become increasingly more valuable for other uses. As well as being the source of energy for the hundreds of millions of cars that populate the planet, hydrocarbons are also an important feedstock for the chemical industry. They are used to make plastics, drugs, and other important accouterments of our technological society. "It will become increasingly clear that it is a crime against the future to take petroleum and burn it," says Columbia University chemistry professor Ronald Breslow, who is also a former president of the American Chemical Society. "This is not just because of global warming, but because we are burning away materials that are tremendously valuable for other uses."

# 10

# Clean Machines: Technology and the Environment

Since the dawn of man, technology, human civilization, and environmental degradation have always gone hand in hand. Now, in the final decade of the twentieth century, the interplay between technology and the environment has become more complex. On one hand, the sheer volume and global reach of machines and the waste products produced by their manufacture and operation pose environmental challenges that some contend threaten life on Earth. On the other hand, several recent trends in technology are actually good news for the planet. One development has been rapid advances in materials science that has resulted in dematerialization—which, in this usage, is defined as using less material while providing the same or better service or function. A vintage 1998 laptop computer, for example, packs as much computing power as a roomful of circa-1980 mainframes. The new generation of machines requires less energy to run, a trend that has been dubbed decarbonization. That translates into a cut in the amount of carbon-based fossil fuel burned to generate that energy, and thus a reduction in polluting gases like carbon dioxide and nitrous oxides. The net result is that by some measures, the air and water in developed countries in North America and Europe are less polluted than they were fifty years ago.

256 • Who Gives a Gigabyte?

There has also been a dramatic transformation in government and industry. From the invention of the first steam engine in 1712 until less than three decades ago, standard industrial practice was to exhaust all combustion fumes up smokestacks, vent all evaporating materials and solvents in the air, and flush all waste liquids and contaminated waste water into sewer systems or natural waterways. Emblematic of that era was a catastrophe in Cleveland, Ohio, in 1969, when the Cuyahoga River carried so much flammable material that it caught fire and destroyed seven bridges. Now the Cuyahoga can support both plant and animal life, and the lands surrounding the river have been redeveloped into natural and recreational sites. On July 4, 1996, an area of the Hudson River north of New York City, long considered a toxic stew of sewage and chemicals, was opened for swimming for the first time in decades. Water-polluting phosphates have disappeared from detergents, and by many measures, the inhabitants of cities like London and New York breathe far cleaner air than did their mid-century forebears.

Such developments as these are leading to a new perspective on the relationship between technology and the environment. Proponents of this new outlook, broadly called *industrial ecology*, contend that the sustainable future of our planet and its burgeoning human population depends on more, rather than less, technology. Advances in materials science, energy efficiency, miniaturization, and information management are seen as indispensable tools to both monitor the condition of natural systems and develop environmentally benign products and services. In its apotheosis, practitioners of industrial ecology envision production relationships between industries in which the waste products of one industry are the feedstock of another, and every molecule that enters a factory leaves as a product. While this is still a vision of the future, many industrial plants in the United States are already practicing some measure of pollution prevention and have dramatically cut their waste emissions.

These positive environmental developments have resulted from a complex mix of factors, including public concerns about the environment and increasingly strict government regulations in developed countries around the world. In the United States, for example, industries face stringent legislation on waste emissions that is expected to become even tighter. In Northern Europe, which is squeezed for landfill space, manufacturers must contend with regulations that require them to dispose of their product packaging and, ultimately, their products. To compete in world markets, the electronics and automotive industries must now meet strict environmental standards, and thus must include environmental considerations at every level of product development. And, to please an increasingly environmentally conscious public, manufacturers strive to create products that earn the imprimatur of international eco-labels.

Such changes notwithstanding, the global community faces environmental challenges of global proportions that are the result of both present and past industrial practices. Worldwide, the emissions and by-products of modern technologies are exacting serious environmental penalties, particularly in the case of air pollution that winds can carry around the world. Cleaning our polluted atmosphere and relieving the long-lived airborne consequences of our industrial age is one of the biggest eco-technological challenges we face. There are also more complex issues that often involve weighing the benefits of a substance to our economy or daily lives against the potential risks that substance may pose to the natural world. Since the 1940s, more than sixty thousand chemical substances have been synthesized, and thousands more are in the pipeline. U.S. industry alone produces 150 million tons of toxic chemicals a year, generating billions of tons of waste, of which about 700 million tons are considered hazardous. In addition, we must still contend with the legacy of past pollution—and ongoing industrial practices—and the unsafe waste dumps that go with them.

Despite laws to control the release of toxic substances, there is actually remarkably little information available on the long-term effects of the release of synthetic chemicals into the environment. In some cases those chemicals can achieve widespread use before their health or environmental implications are suspected. Some examples are polychlorinated biphenyls—PCBs—which, because of their fire-resistant properties, were widely used in the manufacture of electrical transformers and industrial machinery, and which, by various means, have been dispersed. PCBs are among a range of hormone-mimicking chemicals some environmental epidemiologists believe are responsible for damaging the reproductive capacities of a range of animal species, from alligators to humans, and may also be implicated in hormonally induced cancers. Moreover, precisely because they are synthetic, these chemicals persist in the environment because natural processes cannot break them down.

## Unintended Consequences in Action: How a Wonder Chemical Became an Atmospheric Pariah

A striking example of a family of synthetic chemicals whose widespread use has had a global impact are CFCs (chlorofluorocarbons), which within the space of seventy years went from being hailed as wonder chemicals to being decried as atmospheric pariahs because of their role in depleting ozone in the upper atmosphere. The story of their development, use, and current phase-out can be seen as paradigmatic of the ecological principle of "unintended consequences," in which the introduction of a chemical compound, plant, or animal into the environment has an unforeseen effect.

## ABOUT OZONE

The Earth's atmosphere is divided into regions based on the way the temperature changes with altitude. The lowest layer is the *troposphere*, where the temperature drops with increasing altitude. At an altitude of about fifty thousand feet is the *tropopause*, the point at which the temperature starts to reverse. Above is the *stratosphere*, where the temperature rises with elevation. In the troposphere, dense, cold air tends to sink, warm air to rise, and air pollution is carried down to the ground, where it is removed. The stratosphere, by contrast, is stable and virtually cloudless. Air mixes very slowly be-

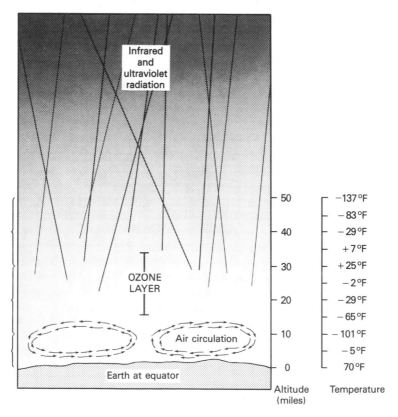

Figure 10.1   The Earth's atmosphere is made up of a series of layers. The troposphere is nearest the Earth. Thirteen to thirty-five miles above that is the stratosphere, where ozone absorbs ultraviolet radiation from the sun.

cause the cooler air is at the bottom. Polluting gases that rise into the stratosphere remain there for years. In the stratosphere, or upper atmosphere, ozone forms a shield that protects life on earth from the destructive effects of ultraviolet (UV) radiation. In the troposphere, ozone, as a product of pollution, is an eye and lung irritant that can cause breathing problems in the elderly and in people with asthma; at very high concentrations, it is a deadly poison.

The increase of ozone in the troposphere and its depletion in the stratosphere are both caused by industrial activity. Ground-level—or atmospheric—ozone is produced when sunlight and heat trigger chemical reactions between gases in the atmosphere and pollutants like volatile organic compounds (VOCs) and nitrogen oxides. VOCs are materials like gasoline, paint solvents, and organic cleaning solutions, which evaporate and enter the air. The reaction that produces ozone also produces other toxic compounds that damage plants and animals. Because of the role of sunlight, these products are known as *photochemical oxidants*; the brown haze that develops is called *photochemical smog*.

## The Ozone Hole

In the stratosphere, a thin layer of ozone protects life on Earth from the destructive effects of ultraviolet (UV) radiation by, in effect, absorbing the energy of that radiation in chemical reactions in which ozone is constantly produced and destroyed. Ozone is depleted when, instead of absorbing ultraviolet radiation in its usual stratospheric oxygen cycle, it gets used up in competing reactions with gases that migrate into the stratosphere from anthropogenic (man-made) sources. The chemicals most seriously implicated in ozone-depleting reactions are halogens, that is, compounds containing chlorine, bromine, fluorine, and iodine, all of which are highly reactive nonmetallic elements. CFCs are the best-known compounds derived from this chemical family, but other chemicals, like halon gas and several insecticides, are also responsible for ozone depletion. In the stratosphere, ultraviolet radiation splits these compounds, releasing free atoms, which then react with stratospheric ozone to form other molecules. Often a second reaction then follows, in which the halogen gas is released to react with yet another ozone molecule. Because these halogen molecules do not get used up in the reaction, they are called *catalysts* and have the potential to destroy numerous ozone molecules in a chain reaction. It has been estimated, for example, that every chlorine atom in the stratosphere has the potential to destroy 100,000 molecules of ozone.

## Tracking Down the Ozone Hole:
## How a Best-selling Chemical Was Banned

CFCs were first synthesized in the late 1920s by a research chemist, Thomas Midgeley Jr., who to demonstrate their safety, sucked in the vapors and then blew out a candle. Stable, noncombustible, and nontoxic, CFCs were used to produce foam products such as bedding, furniture, appliances, packaging, flotation, and fast-food containers. Around the world, millions of room and automotive air conditioners and refrigerators pumped out cold air courtesy of the impressive cooling properties of CFC-12, known more widely by the trademark name Freon. By 1985, world consumption of CFCs was well over 2 billion pounds.

Because of their stability and apparent safety, investigators were hard-pressed to make a connection between life-threatening ozone depletion and a foam takeout container. But the problem lies in the stability of CFCs: gases released in the production of foam products or by leaking refrigerants are virtually indestructible, taking up to ten years to migrate into the stratosphere. There the intense UV radiation causes their disintegration, allowing the released chlorine to attack ozone in a catalytic chain reaction (see above).

In the 1970s, scientists Mario Molina and F. Sherwood Rowland expressed concern, in a letter to the journal *Nature*, that CFCs release chlorine into the stratosphere and that the chlorine destroys ozone. Anxiety about the impact of CFCs increased substantially in subsequent decades with the discovery of a "hole"—or thinning—in the ozone layer over Antarctica. In October 1993, for example, the "hole" covered 9 million square miles and took a 25,000-foot-deep slice out of the stratosphere. Although this springtime thinning, a function of winter Antarctic air circulation and pollution, disappears in the summer, there are concerns that similar reactions can take place elsewhere. There is also evidence of striking springtime ozone depletion in southern hemisphere countries such as Australia and Chile.

The discovery of polar ozone holes and the revelation of thinning ozone elsewhere in the stratosphere galvanized the international community, and in 1987 the United States and twenty-two other countries signed an agreement known as the 1987 Montreal Protocol on Substances that Deplete the Ozone Layer, which called for a phaseout of CFCs. Reports of growing Antarctic ozone depletion and indications of ozone depletion in the northern hemisphere resulted in further revisions that accelerated the complete phaseout of CFCs. In the United States, strict legislation against ozone-depleting chemicals was built into the 1990 Clean Air Act Amendments. Electronics manufacturers, such as IBM, who used CFCs to clean computer chips, completely phased out their CFC use by 1994. On January 1, 1996, production of CFCs in the developed world virtually ended, and production in the developing world is scheduled to end by 2010.

The worldwide compact to phase out the production and use of CFCs is widely seen as an admirable example of global cooperation in the face of a clear environmental threat. It is hoped that this process can be emulated in response to the issues of nuclear waste and global warming, for example. But CFCs already in the atmosphere will be contributing to ozone depletion well into the next century, and CFC manufacturers are still wrestling with the problem of disposing of large stockpiles of this indestructible chemical. There is also the problem of finding CFC replacements, particularly for the world's current stock of air conditioners and refrigerators. It has been estimated that the value of the equipment in the United States that is dependent on CFCs is about $135 billion. Among those affected have been the owners of automobiles equipped with air conditioners that still use Freon. Many face the prospect of a costly retooling when their air conditioners break down. One commonly used substitute, HCFC (hydrochlorofluorocarbon), also contains chlorine, albeit in far smaller quantities, and is a potent greenhouse gas that traps heat 1,200 times more effectively than carbon dioxide. Its use is scheduled to be phased out by 2030. The most promising substitutes are HFCs—hydrofluorocarbons—which contain no chlorine and are judged to have no ozone-depleting potential. But the hydrofluorocarbon of choice—HFC 134A—traps heat 3,200 times more effectively than carbon dioxide. So the molecular search continues.

In Europe, refrigerator manufacturers have turned to hydrocarbons such as butane, which proponents contend are both chlorine free and do not contribute to global warming. In North America, however, manufacturers have resisted this "Greenfreeze" option on the grounds of fire safety and efficiency. Environmentalists have countered that the chemical companies have resisted using hydrocarbons because they cannot be patented, and that they prefer the profit potential in HFCs. There are also concerns about the continuing production and sale of CFCs in the developing world. In 1996 a British researcher claimed that millions of pounds of CFCs were being smuggled into the United States to meet the demands for servicing existing automotive air conditioners. CFCs, he noted, appear to be "the second most lucrative commodity smuggled into the United States through Miami, exceeded in value only by cocaine."

## Life without CFCs

Less than fifteen years ago, CFCs were probably the most widely used synthetic chemicals in the world. One would think that finding a substitute would be difficult, ruinously expensive, and highly disruptive. But alternatives have been found, albeit at a steep price, and the replacement process is instructive of how the global community can address a complex environmental issue. The first impetus for change was persuasive scientific research on the ozone-destructive properties of CFCs, coupled with the conviction that increased ultraviolet radiation would threaten life on earth. The second was that CFCs were produced by a few large manufacturers, such as the Du Pont Corporation, which found the scientific research compelling and acted quickly. The CFC ban has cost industry billions of dollars worldwide. To develop and produce foams using CFC replacements, manufacturers had to invest large sums in fire safety equipment because the substitutes are more flammable. The electronics industry used CFCs to clean components: their chemical inertness ensured minimal surface damage, and their high vapor pressure simplified the process of drying surfaces after cleaning. Finding alternative methods cost hundreds of millions of dollars, and research is ongoing. Investigators are combing chemical databases to find nontoxic substitutes that have the desired properties of CFCs without the deleterious potentials for global warming and ozone depletion. Among the substitutes that this high-tech search has turned up are soapy water and lemon juice!

## Combating Acid Rain

Like the ozone hole, acid rain is a mostly man-made problem with localized causes but global effects. The term is derived from atmospheric studies made in Manchester, England, more than a century ago. The more accurate term for acid rain is acid deposition, because there are other forms of acid precipitation: acid sleet, acid hail, acid frost, and deposits of dry acid particles. Because its mechanisms are less well understood, and its consequences are not as unambiguously deadly as ultraviolet overexposure, worldwide efforts to address the problem have been less forceful, and the industries deemed responsible for the phenomenon have been more resistant to taking action.

The two acids found in acid rain are sulfuric acid ($H_2SO_4$) and nitric acid ($HNO_3$). Burning fuels produce sulfur dioxide and nitrogen oxide gases, which react in the atmosphere with water vapor and hydroxyl radicals (OH) to form sulfuric and nitric acids. Most of the sulfur dioxide is from coal-burning power plants and is often transported long distances in the atmosphere before combining with water vapor to form acid rain. The nitrogen oxides (which also play a role in ground-level ozone creation and stratospheric ozone depletion) are mainly traced to transportation emissions and fuel combustion.

| Alkaline | | | | Neutral | | | | | | | | | Acid |
|---|---|---|---|---|---|---|---|---|---|---|---|---|---|
| 14.0 | 13.0 | 12.0 | 11.0 | 10.0 | 9.0 | 8.0 | 7.0 | 6.0 | 5.0 | 4.0 | 3.0 | 2.0 | 1.0 |
| Lime | Lye | Antacids | Ammonia | Baking Soda | Blood | Milk | Rain | Tomatoes | Vinegar | | | | Battery Acid |

The acidity of a solution is determined by its proportion of hydrogen ions, which is why acidity is referred to as a pH. Acidity is measured on a pH scale of 14, in which a neutral solution has a pH of 7. As the pH decreases, the acidity increases. The scale is logarithmic: water with a pH of 5 is ten times more acidic than water with a pH of 6.

Concerns about the effects of acid deposition on plant life and aquatic ecosystems surfaced about thirty-five years ago, when anglers started noticing precipitous declines in fish populations in many lakes in Sweden, Ontario, and the Adirondack Mountains of upper New York State. Scientists in Sweden were the first to identify the cause as increased acidity of the lake water. Basically, acidity damages the enzymes and hormones of organisms that live in water. The eggs, sperm, and developing young of these organisms are severely stressed, and many die even in small shifts of environmental pH. In addition, there are concerns that acid rain helps to leach toxic heavy metals from the soil and eventually into the water supply.

After extensive study, environmental scientists have blamed acid rain for the "death" of at least 6,500 lakes and seven Atlantic salmon rivers in Norway and Sweden and the disappearance of aquatic life in 1,200 lakes in Ontario, Canada. In the Adirondacks, there are no fish in more than two hundred lakes, and many have no life in them except some algae and bacteria. Although the incidence of acid rain is widespread, the effects of acidification have been somewhat limited because of the buffering capacity of some aquatic systems. The acid in these aquatic systems is buffered, or neutralized, by the presence of limestone ($CaCO_3$), which reacts with the acid to form calcium, carbon dioxide, and water. Around the world, historic buildings and statues constructed of limestone or marble are being steadily eroded by acid rain as a result of the same chemical reaction. Acid rain is also believed to damage susceptible plants. Researchers attribute the decline of red spruce forests in the mountains of northern New York and New England to their immersion in clouds laden with acids and other pollutants that travel hundreds of miles from power plants, refineries, ore smelters, and vehicles in the industrial Midwest.

Recent research has uncovered evidence that the additional burden of toxic heavy metals in polluted air can contribute to the weakening of these trees.

In the United States, environmental laws have addressed the problem of acid rain by reducing the amount of sulfur vented into the atmosphere by coal-burning power plants. To bring about the reduction, the Clean Air Act amendments of 1990 introduced a system of pollution permit trading, which permits utilities to buy and sell the limited number of sulfur dioxide emission allowances, which are priced by the ton. As of 1998, it appears that the trading system has brought about a large reduction in sulfur emissions. The aim of the legislation is to reduce emissions 50 percent by 2000. But acid rain is a growing problem in Asia, especially in China, which uses high-sulfur coal to generate electricity.

## Air Pollution and the Law

Despite massive opposition from a broad array of interests, developed countries around the world are promulgating increasingly stringent environmental regulations to limit and relieve air pollution. Clean air legislation has existed in the United States for more than three decades. The first significant legislation, the Air Quality Act, was passed in 1967. The legislation has been amended several times since then, becoming more comprehensive and strict in the process. In 1990, Congress passed a revision of the Clean Air Act that many consider to be the largest, most complex, and expensive effort in the history of environmental regulation. This legislation is credited with markedly improving the quality of America's air. The tough federal mandates have forced science and industry to produce cleaner cars, cleaner power plants, and cleaner fuels. The legislation is also forcing innovation and the faster development of emission-free technologies, particularly in the areas of power generation and transportation (see chapter 9). Because the legislation has dramatically increased the cost of cleaning up pollution, the problem of industrial emissions is also being addressed by developing more environmentally benign production processes. Research in the newly emerging field of industrial ecology is expanding. In the future, industrial processes will be designed to recycle and reuse polluting gases, rather than vent them into the air. Solvents are a major source of VOCs. Work has already begun on developing stable replacement solvents, and on redesigning chemical processes so that they do not require a solvent at all.

## Air Pollution Control Equipment

Although strict clean-air legislation is of fairly recent vintage, most of the technology to clean gases from such industrial processes as power generation and manufacturing was first developed at the turn of the century. The basic design has remained relatively unchanged. For example, the electrostatic precipitator, an electrical device for removing impurities such as dust, fumes, or mist from air or other gases, was developed by the American chemist Frederick Gardner Cottrell, in 1906. (This kind of precipitator works by "charging" the suspended particles, which are then attracted to a collection plate that has an opposite electrical charge. The collector electrodes are then washed or scraped clean to remove the deposited impurities.) Other methods of controlling particulate emissions include "gravity settlers," which allow particulates to settle out of waste gases as they pass out into the air; mechanical collectors, which force the discharged air stream to change direction and thus precipitate out the particulates; scrubbers and fabric filters, called "baghouses," which are like building-sized vacuum cleaner bags. Because of stringent new regulations regarding venting into the atmosphere and dumping, the emphasis is increasingly on recycling. A recently developed closed system, for example, uses cleanable filters to recycle the lead powder that is sprayed on the inside of television tubes to prevent radiation emissions. The closed filter system captures excess powder and pulses it from the cartridges directly into a recycle system for return to the spraying process, thus avoiding the problem of hazardous dust and cartridge disposal.

## Wallboard from Air Pollution

Scrubbers are most widely used in the power industry to remove the sulfur dioxide that is discharged when coal is burned to generate electricity. Scrubbers are "liquid filters" that remove the sulfur dioxide by converting it into a harmless solid. For example, in a venturi scrubber, exhaust fumes pass through a spray of water that contains lime. The sulfur dioxide reacts with the lime and is precipitated as calcium sulfate ($CaSO_4$). The pollution/limestone reaction produces a sludge that is usually dumped on the ground or into pits or waterways. This creates a new waste disposal problem: a typical 500-megawatt power plant burning high-sulfur coal produces about 10 million cubic feet of gypsum sludge a year, enough to cover a football field with a mound of gypsum over two hundred feet high. In Germany, where all power plants are equipped with pollution controls, the law prohibits sludge dumping. So the Germans devised a better way to use the scrubber waste: in the production of gypsum for wallboard. The process was actually developed in the United States in 1973 and exported to Germany in 1980 because in the United States at that time, utilities were allowed to build tall stacks at their generating plants for dispersing sulfur dioxide over wide areas.

## Air Pollution on the Move

"Mobile sources"—that is, cars, trucks, buses, and airplanes—release nearly half of the pollutants that foul our air, including volatile organic compounds, carbon monoxide and the nitrogen oxides, that lead to acid rain and ground-level ozone. (In the mountain forests east of Los Angeles, which has automobiles in profusion, scientists have found fog water of pH 2.8—almost a thousand times more acidic than usual—dripping from pine needles.)

Because of technological advances in pollution control, the current generation of cars is far less polluting than its predecessors. The fuel economy of new U.S. cars doubled between 1974 and 1985, and the addition of electronic engine controls and catalytic converters to lower hydrocarbon and carbon monoxide output has cut pollutants by more than 75 percent since 1970. Since the 1990 passage of the Clean Air Act amendments, automobile manufacturers have reduced polluting emissions even further. Ford Motor Company, for example, which is marketing a range of alternatively fueled vehicles, claims to have reduced hydrocarbon emissions in its fleet by more than 98 percent since the 1960s. Computers control the fuel mixture and ignition timing, which allows fuel to burn more completely. This decreases VOC emissions. However, the technology that has been most responsible for cutting automobile emissions is the catalytic converter. As exhaust gases pass through this device, a chemical catalyst made of platinum-coated beads oxidizes most of the VOCs to carbon dioxide and water. The catalytic converter also oxidizes most of the carbon monoxide to carbon dioxide.

But this gain has been offset by the explosive increase in the number of cars. In the United States, for example, between 1970 and 1990, the number of miles traveled in cars doubled, from 1 trillion to 2 trillion miles per year. According to the Worldwatch Institute, an environmental monitoring organization, the world's car population presently stands at 501 million, and is expected to reach the 1 billion mark by 2020. Environmentalists are also concerned that the burgeoning demand for cars in developing nations will place additional burdens on fragile environments as valuable agricultural land is paved to build roads and parking garages.

In the United States the problem of air pollution from "mobile sources" has been addressed by the strict legislation of the amended Clean Air Act. For example, beginning in 1995, special "clean" fuels, treated with grain alcohol, have been required in cities with the worst air pollution. Several large cities are experimenting with cars and buses powered by compressed natural gas (CNG), which consists mostly of methane and can be significantly less polluting than diesel fuel or gasoline. Indeed, automotive engineers cite emission regulations as the top technological challenge they face now and in the future. One approach is to improve monitoring and feedback systems so that fuel is burned more completely and exhaust gases are catalyzed more efficiently. These systems require the use of sophisticated sensors. New emissions control systems will include catalysts and sensors that are electronically heated so that they can measure the number of individual exhaust gas constituents within seconds after the engine is started. Because catalysts don't work at low temperatures, most emissions occur during the first few minutes of car operation. Another, more radical approach to nonpolluting transportation, which is discussed more fully in chapter 9, is the development of alternatives to the internal combustion engine so that the inevitable air pollution that results from the combustion of fossil fuels can be avoided.

## Land and Water: From Toxic Dumps to Pollution Prevention

The cornerstones of environmental legislation in the United States were the Clean Air Act of 1970 and the Clean Water Act of 1972. But the laws left an enormous unregulated loophole: companies were free to dump their wastes on land. As problems surfaced—and fires at several toxic waste dumps riveted public attention—Congress, in 1976, passed the Resource Conservation and Recovery Act (RCRA). The main features of this act required that all disposal facilities such as landfills have permits, that these landfills possess such safety features as monitoring wells, and that wastes destined for landfills be pretreated so that they would not leach. The law also required that hazardous wastes be tracked from "cradle to grave."

But the legacy of unregulated land disposal and illegal toxic waste dumping remains in thousands of abandoned toxic waste sites all over the United States—some in urban centers and others in surprisingly remote rural areas. In 1980 the U.S. government launched a huge program to clean up chemical waste sites with the Comprehensive Environmental Response, Compensation and Liability Act, known more popularly as Superfund. This legislation provides funding to identify abandoned chemical waste sites, protect the groundwater near the site, and clean up the site. To date, more than 1,300 sites have been placed on the National Priorities List for urgent remediation. As of February 1998, cleanup and construction on 509 sites had been completed.

## HAZMATS and Why They Matter

HAZMATS—the U.S. Environmental Protection Agency term for hazardous materials—are chemicals that present a certain hazard or risk. The EPA categorizes substances based on whether they have dangerous properties like chemical instability and toxicity. Also of concern are chemicals that catch fire readily, corrode storage tanks and equipment, or are radioactive. Natural processes can gradually break down and assimilate many of the chemicals that are introduced into the environment. Dispersing those chemicals so that nature can take its course is known as the "dilution solution." However, natural processes cannot take care of HAZMATS such as heavy metals and their compounds and nonbiodegradable synthetic organics, like PCBs. The most dangerous heavy metals are lead, mercury, arsenic, cadmium, tin, chromium, zinc, and copper. As ions or in certain compounds, heavy metals dissolve in water and are absorbed into the body, where they tend to combine with and damage vital enzymes.

Synthetic organic compounds are materials that owe their provenance to the ingenuity of chemists. They are the basis for all plastics, synthetic fibers, synthetic rubber, solvents, pesticides, and hundreds of other products. Nonbiodegradability is an important part of what makes many such compounds useful. But these compounds are toxic because they are similar enough to natural organic compounds to be absorbed into the body. For example, most mammalian cells have receptors that bind to dioxin, PCBs, and many other toxic pollutants. Once attached to the cell, these complexes travel into the cell's nucleus, where they can inflict genetic damage.

## Biomagnification and Bioaccumulation

Heavy metals and nonbiodegradable synthetic organics are particularly hazardous because they tend to accumulate in living things so that over time small, seemingly harmless amounts received may reach toxic levels. Traces of heavy metals and synthetic organics that are absorbed with food and water are trapped and held by the body's enzymes and fatty tissues. Because the body cannot excrete heavy metals and synthetic organics or break them down, trace levels consumed over time gradually accumulate. Bioaccumulation may be compounded through a food chain. Each organism accumulates the contamination in its food. The contaminants eaten by organisms like algae, at the bottom of the food pyramid, are concentrated, through food chains, into the systems of animals at the top of the food pyramid. This multiplying effect is called *biomagnification*. The tragic consequences of these processes have been starkly demonstrated in the fate of a small population of beluga whales in the heavily polluted St. Lawrence River in Quebec. The whales, which feast on contaminated krill and fish in the St. Lawrence, suffer disproportionately from cancer and widespread reproductive problems.

## SUPERFUND AT WORK:
## THE MARATHON BATTERY COMPANY

Environmental remediation is a laborious and hugely expensive process: environmental cleanup technology has relatively few methods at its disposal, and there are no high-tech quick fixes. The process is illustrated in the cleanup of a 350-acre contaminated site known as the Marathon Battery Company in Cold Spring, a historic village on the Hudson River about sixty miles north of New York City.

For almost thirty years, nickel-cadmium batteries were manufactured at this site, and between 1952 and 1965, the plant discharged untreated wastewater into the Hudson River and a marsh. State officials were alerted in the mid-1970s, when a canoeist noticed a green fluid trickling out of a pipe into the marsh. Tests revealed extraordinarily high levels of heavy metals, including cadmium, nickel, and cobalt at the plant site; in river sediments; and in the marsh soil. Tides had flushed cadmium deposits from the wetlands into the Hudson River. Area groundwater also contained elevated levels of trichloroethylene (TCE), a widely used solvent. In the marsh, the muskrat population had plummeted 85 percent because of cadmium contamination in marsh cattails, their primary food source. There were also concerns about the contamination of fish, blue crabs, and numerous plant species in the Hudson River. The area was declared a Superfund site in 1983.

To clean up the marsh and the Hudson River, 95 percent of the sediments were dredged and chemically stabilized. This involved building a dike to drain the marsh, drying the soil, and then mixing it with a chemical compound designed to transform the cadmium and other heavy metals into nonsoluble salts. The contaminated marsh was covered with an impermeable clay cap, called a Bentomat, which is a carpet of clay sandwiched between two pieces of strong fabric. New soil was trucked in, and sixty thousand cattails and other marsh grasses replanted.

In all, 189,265 tons of treated soils and sediments were transported via 1,979 railroad cars to a regulated landfill in Michigan. A hazardous waste landfill almost five hundred miles away, in upstate New York, received 906 tons of untreatable materials from the same site. The cleanup cost $91 million and took 2½ years.

## New Cleanup Methods

Future trends in cleanup technologies may include more use of *in situ* remediation, where the pollutant is treated on site rather than being trucked away. Among cleanup processes that are being developed and may play a significant role in the future is an electrokinetic treatment method in which electrodes are placed in the soil. This causes the contaminants to migrate to either the positive or negative electrode. In some cases the pollutant is then treated in a special container that is located near the electrode.

### LEACHING AND GROUNDWATER CONTAMINATION

Because of the dangers they pose to living things, one of the main concerns of environmental remediation is to prevent leaching—the process in which hazardous materials in or on the soil gradually dissolve and are carried by water seeping through the soil into groundwater. Worldwide, a large proportion of drinking water is from groundwater—or aquifers—and chemical contamination is a very serious concern. A spreading chemical in groundwater is called a plume. "A very small percentage of the water in the world is drinkable," noted an EPA engineer. "An unrestrained plume will keep on spreading, and if you lose an aquifer as a source of water, it's pretty much lost forever."

There are an estimated 5 million to 6 million underground storage tanks in the United States that contain either a hazardous substance or petroleum. Of those, approximately 400,000 are believed to be leaking, with many more at risk. Substances released from leaking tanks can poison crops, damage sewer lines and buried cables, and lead to fires and explosions. The most serious concern, however, is groundwater contamination. One gallon of gasoline is enough to render a million gallons of groundwater unusable, based on federal drinking water standards. Leaching prevention is also one of the major focuses of a modern municipal landfill. No longer a smoky pit near town, municipal garbage disposal is becoming increasingly technologically advanced.

# Back to Nature: The Miniature World of Bioremediation

One alternative environmental technology that could play a significant role in the future is bioremediation: the use of living organisms, mainly microorganisms like bacteria, to break down environmental pollutants or to prevent pollution through waste treatment. Microorganisms have an enormous capacity to degrade organic compounds like fuel. Researchers looking to harness this potential are searching for microorganisms that can grow under extreme environmental conditions, amid high concentrations of solvents, or in high temperatures or extreme alkalinity. And if these microorganisms do not occur naturally, or they cannot degrade a synthetic substance, genetic engineers believe that they will be able to construct them (see chapter 5). One research group in Japan, for example, has isolated a strain of *Pseudomonas* bacteria that can grow in solvents containing more than 50 percent toluene, conditions that kill most organisms. Toluene is a flammable liquid that is used in aviation fuel, explosives, solvents, and dyes.

Microorganisms are already hard at work in waste treatment systems in Europe. In the Netherlands, which, with Germany, is considered a leader in eco-technology, biofilters are used to remove organic contaminants from the air. In a waste treatment application, a dry-composting process converts biodegradable organic solid waste and refuse into energy in the form of biogas (methane and carbon dioxide) and a humuslike material that can be used for compost. The biogas is produced by a colony of various anaerobic bacteria (which do not use oxygen) that include methanogens (methane-producing archaebacteria). Bacteria are also used to remove nitrates from wastewater, thus preventing waterway eutrification. (Eutrified water is so nutrient-rich that it supports an overabundance of algae and other aquatic plants at the water's surface, cutting off oxygen to the deeper water. Fish and any deep-water plant life are killed as a result.) New waste treatment systems use microorganisms that can biodegrade such compounds as hydrocarbons and chlorinated solvents found in industrial plant wastewater. In bioscrubbers and biotrickling filters, multiple microbial communities grow on solid surfaces to produce multilayered complexes called biofilms. When gases containing organic pollutants are passed through these systems, the pollutants are degraded.

Another focus of bioremediation research is to address future environmental problems. In Japan, researchers are studying ways to harness microbes to produce substances that can replace petrochemicals for energy generation and industrial production. They are also experimenting with using microorganisms to produce biodegradable plastics and remove sulfur from industrial air emissions. Japanese researchers are trying to develop microorganisms that can help reverse desert formation, and a bioremediation system that can remove carbon dioxide from the atmosphere (see chapter 9). One laboratory has isolated algae that convert carbon dioxide to carbohydrates ten times faster than a terrestrial green plant.

In the United States, bioremediation is mostly used to clean up sites contaminated by toxic chemical spills, chemical waste disposal, and leaking underground storage tanks. Biological treatment is at least ten times cheaper than moving and incinerating large quantities of polluted materials. A highly successful approach to bioremediation is to stimulate the activity of naturally occurring organisms. To clean up the 1989 *Exxon Valdez* oil spill, in which 11 million gallons of crude oil were spilled into the pristine Prince William Sound in Alaska, crews first tried using high-pressure water to wash the rocks. Bioremediation, which in this case consisted simply of adding nitrogen-containing fertilizers to the contaminated shorelines, was more successful. It stimulated the metabolism of indigenous hydrocarbon-degrading microorganisms and degraded both surface and subsurface oil three to five times faster than at untreated test sites. The cost of remediating hundreds of miles of contaminated shoreline was less than $1 million, although the tests and studies that preceded this cost considerably more. Naturally occurring microorganisms have also been used to clean contaminated groundwater. In Pennsylvania, researchers successfully used bioremediation to remove hydrocarbons from a gasoline-contaminated aquifer. In a process called *bioventing*, the hydrocarbons were removed by adding fertilizers and using air bubbles to stimulate the growth of indigenous hydrogen-utilizing microorganisms in the groundwater. Recently, scientists at Lawrence Livermore National Laboratory successfully used bacteria to break down trichloroethylene (TCE) contamination in groundwater 30 meters below the surface.

Bioremediation promises to be cost-effective and potentially "cleaner" than conventional cleanup methods. The method, however, is not without problems. Every site has different characteristics, which can sometimes be quite subtle; therefore, detailed and expensive studies of both the nature of the site and the chemicals present are required. Microorganisms degrade only very specific classes of chemicals, and pollutants are often mixtures of chemicals. Crude oil, for example, contains thousands of hydrocarbons with different chemical structures. Rates of microbial activity are also influenced by temperature and other environmental conditions, and bioremediation can be slow. There are also concerns about maintaining microbial activity when the organisms have degraded most, but not all, of their target chemical. Some compounds, such as TCE, are only biodegradable through co-metabolism with another substance. In some cases the added chemical may also be toxic.

One of the biggest problems in bioremediation is measuring microbial activity. One novel method of monitoring the presence and activity of specific microorganisms uses "reporter genes," which produce an easily monitored effect when microbial activities are occurring. Gary S. Sayler at the University of Tennessee, Knoxville, uses bioluminescence to monitor the degradation of naphthalene, a component of crude petroleum, in soil. To do so, he constructed genetically engineered microorganisms in which a gene coding for bioluminescence is fused to the genes of a species of bacteria that degrades naphthalene. As the engineered bacteria degrade the naphthalene, a measurable amount of light is emitted. Sayler has proposed using reporter bacteria as biosensors for *in situ* environmental monitoring. Working with researchers at Oak Ridge National Laboratories, he has combined reporter bacteria with an integrated circuit to produce a half-living, half-silicon chip in which the light emitted by the bacteria is translated into an electrical signal. This "critter on a chip" could replace today's complicated and expensive optical pollution detectors.

## Phytoremediation

Another developing method of harnessing Mother Nature for environmental cleanup is *phytoremediation*, or the use of green plants. Certain plants, such as specific strains of *Brassica* (mustards), accumulate heavy metals when growing in metal-contaminated soils. These plants can accumulate up to 40 percent of their biomass as heavy metals. They can then be harvested and the metals recovered for recycling or disposal. A variation of this strategy is to use plants that exude a chemical that promotes bioremediation. Because poplar trees have demonstrated these properties, they have been planted on top of landfills.

## Green Technology: Some Key Concepts

In the complex interplay of technology and the environment, a new school of environmental thinking is developing in which technology is considered a potential ally rather than the enemy. This thinking contends that we will need more and not less sophisticated technology to sustain our natural environment as well as meet the economic needs of the world's burgeoning population. But this is not "technology-as-usual"; rather, it proposes using sophisticated technologies and industrial processes to produce goods and services in an environmentally benign way. Instead of concentrating on environmental cleanup, these practitioners believe that our technological expertise has advanced to the point where we do not need to pollute. Some of the concepts and vocabulary of this field are detailed below.

**deep ecology**   An approach to sustainable development that advocates limitations on the use of technology. (A deep ecologist will use an abacus and ride a bicycle.)

**design for disassembly**   The practice of designing a product so that its components can easily be broken down and recycled.

**design for the environment (DFE)**   An approach to product design that incorporates environmental concerns and pollution prevention from the design stage to the end of the product's life cycle.

**eco-development**   This paradigm gives equal weight to human society and ecosystems. The economy relies principally on renewable sources of energy and materials, extracted at rates that would not affect ecological health. Nonrenewable resources would be recovered and recycled indefinitely.

**green design**   A design process in which environmental attributes, such as waste prevention and better materials management, are treated as design objectives rather than as constraints. Green design avoids using materials that are toxic to humans or ecological systems, substitutes renewable for nonrenewable materials, and ensures that nonrenewable materials can be readily recovered for recycling.

**green products**   Products whose manufacture, use, and disposal place a reduced burden on the environment.

**industrial ecology**   The study of the relationships among firms in industrial production networks and of the flow of energy and materials within the network, the wider economy, and the natural world. (The industrial ecologist would use a laptop computer and drive an electric car.)

**industrial symbiosis**   In the natural world, organisms rely on each other for nutrients and absorb each other's wastes: Companies practicing industrial symbiosis exchange by-products for use as feedstocks rather than continually using virgin materials and discarding waste.

**life cycle analysis (LCA)**   This analysis takes an individual product and maps the flow of energy and raw materials that were required to create the product through its entire life cycle.

**sustainable development**   Development that meets present economic and societal needs without compromising the ability of future generations to meet their needs.

Traditional waste management strategies concentrate on "end-of-pipe" pollution controls, such as the catalytic converter at the exhaust pipe of an automobile. But with stricter environmental regulations, this option is becoming increasingly costly. It was estimated in 1994, for example, that the U.S. chemical industry spends $4.4 billion annually to control, treat, and dispose of the billions of pounds of hazardous waste it produces. New environmental thinking, however, contends that factors like the enormous expense of cleanup, increasingly stringent environmental and waste management regulations worldwide, and environmental concerns on the part of the public are pointing the way toward pollution prevention. Rather than treating wastes and, in effect, transferring the contaminant from one medium to another, this strategy attempts to handle waste at the source, recycling the wastes or redesigning chemical processes and products so that wastes are prevented or put to productive use. Pollution prevention, therefore, in the example cited above, would involve redesigning an automobile engine so that it produces less pollution, or switching to an electric car. Many environmental scientists contend that pollution prevention will be the "environmental option" of the twenty-first century.

Although pollution prevention is by no means universal, emissions reductions and the substitution of less toxic components are already developing in a broad array of industries. Employing design for the environment principles, for example, the Ford Motor Company redesigned a heat exchanger to eliminate the use of TCE, a hazardous industrial solvent. In the 1970s, the heat exchanger was made of copper-brass and silver, with a lead solder, and was cleaned with TCE. Now the heat exchanger is made of an aluminum alloy that is cleaned with water and a detergent. Industrial chemists are substituting more benign processes for the toxic methods that were previously used. Gold sulfides are replacing cyanides in some metal plating baths, vegetable-derived compounds are supplanting chromium in leather tanning, and carbon dioxide is replacing phosgene in the manufacture of polyurethane. Chemical engineers are also rethinking process design, like designing out the need to store a toxic chemical. For example, the methyl-isocyanate (MIC) that killed thousands in Bhopal, India, in 1984, was released from a storage vessel. Now MIC, a chemical intermediate in the manufacture of an agricultural chemical, is created on demand and consumed in the process.

Research chemists are also looking at harnessing natural processes to produce industrially important chemicals. In one experimental program, chemists used genetically engineered *E. coli* bacteria to produce catechol, a feedstock for vanilla flavoring, and adipic acid, a key ingredient in nylon manufacturing. Production of both chemicals would otherwise require benzene, which is considered a carcinogen, as a starting material.

In the United States, the paper industry provides a striking illustration of developing pollution-prevention principles. Twenty years ago the typical American paper mill spewed millions of gallons of contaminated water into a nearby river or stream and belched rank fumes that corroded the metal of cars for miles around. Now many mills have cut their waste water by two thirds, ended most visible pollution, and reduced invisible emissions by up to 90 percent. In contrast to previous practices, water and chemicals are reused. The water, chemicals, and lignin (a by-product) are heated in a recovery boiler. Water is driven off as steam, the wood residue is used as fuel. The chemicals are purified and mostly reused. In some advanced plants, the water and chlorine dioxide used to whiten the paper are sent to a treatment tank. The chemicals are removed and the pure water returns to the bleacher for reuse. Many of America's mills are heading toward "closed loop" systems that eliminate the discharge of water altogether. But some paper industry executives caution that there are chemical problems that must be solved before 100 percent recycling becomes a reality. Developments in de-inking and papermaking technology have also revolutionized paper recycling. New methods of pulping and crisscrossing fibers have improved quality to the point where it is hard to tell the difference between recycled and "virgin" paper. According to the Worldwatch Institute, the global consumption of recovered paper is now increasing at a faster rate than that of wood pulp.

## Clean Production: Design for Recycling

A natural ecosystem disposes of its wastes and replenishes its nutrients by recycling all of its elements. In contrast, our human systems are largely based on the unidirectional flow of elements, and, at present, despite glimmers of change, trends in the United States and other industrialized societies tend toward more waste rather than less. There are estimates that 55 million obsolete computers and at least 30 million telephones will end their useful lives in landfills by the year 2005. For various reasons, including liability concerns, many high-tech products are neither designed nor priced to be repaired. And advances in materials science, while delivering considerable benefits, such as lighter cars and tamper-resistant snack bags, have also made many products more difficult to recycle. Nor do present economics, in which, for example, the extraction of virgin materials in the United States and other countries is government-subsidized, favor recycling. By some estimates, recycling programs in place in many cities in the United States actually cost city governments more than would comparable landfill space. In fact, at present, the only materials whose recycling makes economic sense are paper, aluminum, and various heavy metals.

In Europe and Japan, however, landfill space is rapidly shrinking, and many countries, most notably Germany, are mandating that manufacturers of products ranging from automobiles to toothpaste tubes take responsibility for their waste and ultimately for the entire life cycle of their products. While there are still vast reaches of potential landfill space in the United States, manufacturers are also developing new approaches to product recycling. They, too, face the demands of meeting the standards of international markets, stricter waste disposal regulations, and the expectation that waste reduction will eventually become an imperative in the United States as well. Issues American manufacturers are starting to contend with include consumer protection and the implications for quality of using recycled materials. There are also logistical obstacles, such as identifying the components of a given device, what hazardous materials they contain, and which are salvageable or contain precious metals.

Heavy-metal recycling delivers considerable environmental benefits. In addition to avoiding the environmental damage wreaked by mining activities, it has been estimated that recycling one ton of steel, for example, saves 2,500 pounds of iron ore, one thousand pounds of coal, and forty pounds of limestone while using two-thirds less energy. But worldwide, the growth of steel recycling may be slowed by impurities that make recycled materials brittle. To design for recycling, the automobile of the future may be built without copper and other elements that hinder recycling.

To increase the potential to recycle electronic products, a collaboration of European-based companies is developing a standard communications protocol for providing reuse information. They envision embedding tags in each piece of gear that would store critical takeback data about its components. Other approaches, such as chemical markers or bar codes, are being explored. One California company is developing rapid-identification equipment for plastics, based on infrared spectroscopy. Depending on the polymer's chemical structure, the reflected infrared energy varies in intensity at various wavelengths. The data is matched against the spectra of known materials.

With these factors in mind, eco-technologists envision that future products designed for recycling will have small numbers of different materials and individual components. Use of toxic materials will be avoided, and materials will be joined in such a way as to make them easy to separate. In the future it is possible that companies will not even sell products, but rather provide services on products that they continue to own and periodically upgrade. An early example of this orientation is the Xerox Corporation in the United States, which remanufactures many used parts from its copiers, and has changed the way they are manufactured to facilitate this. For example, parts that used to be welded together are now fastened, making them easier to take apart. Other large American corporations that are remanufacturing products include Lucent Technologies, Dupont, and the Saturn division of General Motors. Proponents of this approach contend that selling a service rather than a product results in far less waste and environmental damage. They cite as an example the result when a pesticide company undertook to control pests rather than sell pesticides. The outcome was a substantial reduction in pesticide use.

## TOWARD THE RECYCLABLE ROADSTER

Cars are already one of the most highly recycled products in the United States. When an old car is junked, it is often first sent to a dismantler, who removes any salable parts. The hulk is then crushed and sent to a shredder, where metals are recovered. At present, about 75 percent of materials in old automobiles are recovered and recycled. The remaining 25 percent, consisting of plastics, rubber, and other substances, is generally landfilled. In Europe, the landfilling of old automobile hulks is a growing problem. German car manufacturers face legislation requiring them to take back and recycle old automobiles at the end of their lifetime. BMW, for example, has already introduced a two-seat roadster model with plastic body panels designed for disassembly and labeled as to resin type for recycling. The goal of the BMW facility is to learn to make an automobile out of 100 percent reusable/recyclable parts by the year 2000.

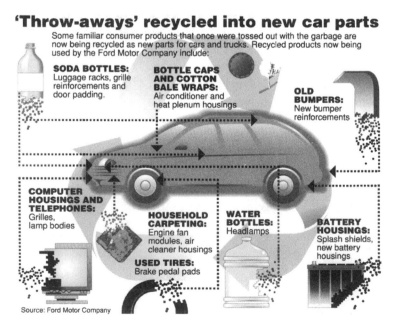

**'Throw-aways' recycled into new car parts**

Some familiar consumer products that once were tossed out with the garbage are now being recycled as new parts for cars and trucks. Recycled products now being used by the Ford Motor Company include:

**SODA BOTTLES:** Luggage racks, grille reinforcements and door padding.

**BOTTLE CAPS AND COTTON BALE WRAPS:** Air conditioner and heat plenum housings

**OLD BUMPERS:** New bumper reinforcements

**COMPUTER HOUSINGS AND TELEPHONES:** Grilles, lamp bodies

**HOUSEHOLD CARPETING:** Engine fan modules, air cleaner housings

**USED TIRES:** Brake pedal pads

**WATER BOTTLES:** Headlamps

**BATTERY HOUSINGS:** Splash shields, new battery housings

Source: Ford Motor Company

**Figure 10.2** Cars are the most extensively recycled products in the United States. In addition, manufacturers like the Ford Motor Company are recycling other consumer products to manufacture new parts for cars and trucks.

## Industrial Ecology: High Tech and Sustainable Development

Many of the precepts of pollution prevention are articulated in the developing field of industrial ecology. Like deep ecologists, industrial ecologists believe that at all levels of our society and economy it is necessary to take environmental considerations and constraints into account. But they differ from deep ecologists in their view of the role of technology in the transition to a sustainable world. The deep ecologist cherishes a vision of a simple agrarian world; the industrial ecologist believes that a sustainable world that can support the burgeoning global population will need to be even more complex, requiring more, rather than less, technology and electronics. No stone-age throwbacks, industrial ecologists assemble comprehensive databases and computer models of the regional and global environmental effects of human activities. The practice includes intensive R&D on environmentally responsible energy supply and use, materials science, and manufacturing technologies. Rather than conventional waste treatments, in which waste is basically shifted around, industrial ecologists strive to reduce the amount of waste and substitute benign chemical processes for toxic ones. Their vision holds that, like natural organisms that depend on each other for nutrients and absorb each other's waste, companies should practice industrial symbiosis, exchanging by-products for feedstocks rather than continually using virgin materials and discarding waste. The ultimate goal is to imitate natural ecosystems, in which there is no waste because mass is completely recycled.

Although elements are already in process, practitioners view the full implementation of industrial ecology as a practice of the future that must await product reformulations, raw material substitution, and significant cultural, political and social change. For the present, the strategies available to the industrial ecologist are improved maintenance and operating practices and relatively simple process modifications that employ currently available technology. Many observers believe that the most difficult task for industrial ecology will not involve developing the technology but changing the social and economic systems that sustain current industrial practices.

### A DANISH SEASIDE TOWN PRACTICES INDUSTRIAL SYMBIOSIS

Officials nurturing plans for industrial parks that follow industrial ecology principles travel to the Danish seaside town of Kalundborg to see their dreams in action. Over the past twenty-five years a complex industrial symbiosis has developed there. The town's coal-fired power station distributes about 224,000 tons of excess steam per year to heat almost all of the town's five

thousand houses and buildings, as well as to provide steam for heat and processing to a large oil refinery and to Novo Nordisk, a manufacturer of pharmaceuticals and enzymes. The power plant also uses excess heat to run its
own fish farm, and sells the sludge from the ponds as fertilizer. A two-mile
pipe runs along a main road to connect the three plants. The power station
supplies calcium sulfate, a by-product of its sulfur dioxide scrubber, to a neighboring wallboard maker, and sells 200,000 tons of fly ash and clinker, residues
of coal burning, for building roads and producing cement. The refinery pumps
its residue gases to the wallboard maker to fire its drying ovens, and sells the
sulfur removed from the gas to a sulfuric acid producer. To conserve fresh
water, the oil refinery pipes cooling water to the power plant, which uses it as
boiler feedwater and to clean industrial equipment. Meanwhile, Novo
Nordisk, a world leader in the manufacture of insulin, enzymes, and peni-
cillin, sells its nutrient-rich sludge to a thousand nearby farms for use as fertilizer. The company uses steam from the power plant to sterilize the sludge.
Distributing the sludge as fertilizer was a way to comply with regulations imposed in 1976 that prohibited companies from discharging sludge into the sea.

## Life-Cycle Analysis

As previously noted, evaluating the environmental impact of a product can
be problematic; the lighting element of an energy-efficient fluorescent lightbulb is mercury, a toxic heavy metal. Semiconducting materials like gallium
arsenide, which offer great functional—and potentially environmental—
benefits are made from equally toxic substances. Life-cycle analysis, which
starts with a particular product and identifies all of the precursors to the product's manufacture, use, and disposal, is a tool used by manufacturers and designers to weigh the trade-offs in environmentally responsible decisions. The
essence of life-cycle assessment is the evaluation of the environmental, economic, and technological implications of a material, process, or product
across its life span from creation to waste. The results of a careful life-cycle
analysis are often counterintuitive. For example, if one factors in the amount
of energy consumed to wash and dry a knit polyester blouse, the environmentally correct result would be to buy a new blouse after every four washings.

## Life-Cycle Analysis Looks at the Paper Cup

Applying life-cycle analysis to the much-maligned polystyrene foam hot-
drink cup has also produced an unexpected result. A frequently cited study

of the relative merits of paper versus polystyrene foam for hot-drink containers by Martin Hocking of the University of British Columbia found that manufacturing the paper cup required more raw material and consumed as much petroleum as the polystyrene vessel. He also found that the papermaking process used more chemicals and wasted more water; the only significant emission associated with the polystyrene cup, he noted, is the pentane used as a blowing agent. The poly cup is easier to recycle and approximately as easy to incinerate. Although theoretically the paper cup is more degradable, recent studies indicate that even "biodegradable" materials remain undegraded in anaerobic landfills over very long time periods, rendering that advantage relatively unimportant. Interestingly, a Dutch study suggested that both options are preferable to ceramic reusable cups, unless the latter are washed hundreds of times before being discarded.

## THE ELECTRONIC INDUSTRY GOES EC

In the opinion of many observers, the electronics industry embodies the ambiguities of twenty-first-century environmentalism: clean technologies are dependent on electronics for process control, monitoring, and information management. But the manufacture of complex electronic instruments ranging from computers to semiconductors involves enormous amounts of water and the use of highly toxic chemicals.

The electronics industry has been striving for environmental correctness—phasing out, for example, the use of ozone-depleting CFCs in 1994. In 1996 the International Standards Organization (ISO), the international electronics forum that sets electronics standards such as telecommunications protocols, agreed on an International Environmental Management Standard. (The standard is called ISO 14000; other environmental standards are similarly identified.) Companies wanting to compete in world markets will have to be certified as meeting certain minimum conditions. Several organizations are also developing sophisticated databases to help electronics engineers make environmentally benign design decisions. In the United States, for example, the Microelectronics and Computer Technology Corporation is developing a sophisticated environmental database that it calls the Environmental Information Mall. The mall will incorporate intelligent-agent software, neural networks, and other high-level information technologies into a comprehensive source of environmental information for the electronics industry.

Addressing the environmental challenges of electronics manufacture is a complex problem because numerous suppliers and manufacturers are involved. Often small shops handle the most toxic processes. Nevertheless, some solutions are being developed. For example, the problem of toxic gases is being addressed by point-of-use-generation, in which substances are kept in their least dangerous form for as long as possible. An alternative to volatile solvents such as acetone and isopropyl alcohol is to use dry-cleaning technology, in which a very cold inert gas—argon, for example—is dispensed through a nozzle aimed at the unclean surface. This expanding gas causes a further temperature drop and a shift into a solid state so that clusters of atoms traveling at very high speed bombard the surface, washing away residues and particles.

The electronic industry's gluttonous appetite for water—up to 15,000 liters of which are used to clean chips for one wafer—is also being addressed. Strategies for decreasing water consumption include redesigning rinse tanks and refining rinsing technology, as well as switching to waterless cleaning methods and reusing the water for other applications. To meet the industry's need for pure water, the Lawrence Livermore National Laboratory has begun to license a cleaning process that uses a carbon aerogel. The porous aerogel, which has an enormous surface area, acts as a large electrode that attracts and thus filters out impurities.

## The Ultimate Greenhouse

Thirty miles from Tucson, Arizona, a huge steel-and-glass edifice named Biosphere 2 sits in the Sonoran Desert. It is a sealed "greenhouse," designed to be self-sustaining, and it is almost completely cut off from the outside world. Built at a cost of $150 million, the 3.1-acre Biosphere 2 is the largest totally enclosed ecosystem ever constructed. Its development and history provide a salutary lesson on the relationship between technology and the environment.

Since "Biosphere 1" is the Earth, many regard Biosphere 2 as one of the most grandiose exercises in eco-technology ever attempted. To create the various ecological systems, nearly four thousand species of plants and animals were gathered from around the world. Fifty miles of sealing glue (taste-tested to be unsavory to termites) anchor thousands of panes of glass. A vast "technosphere"—two hundred motors, fifty miles of pipes, and uncounted miles of cables and hoses—helps maintain the environmental systems, which include algae scrubbers, a tide machine, and a huge "lung" or air-exchange system. A natural-gas-fired electrical plant powers an air-circulation system; 1,600 sensors monitor temperature, relative humidity, light levels, soil moisture, and $CO_2$. A sophisticated circulation system cycles water through the various ecological zones. Vapor from evaporation and transpiration of plants is condensed and circulated

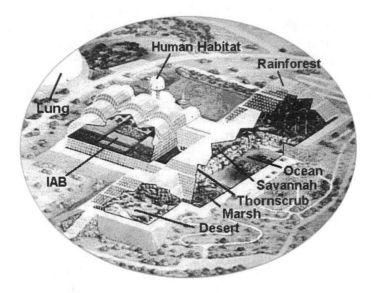

**Figure 10.3** Biosphere 2, possibly the world's largest greenhouse, is divided into seven biomes, or climatic regions, which are characterized by specific plants and animals. Among the regions is an intensive agricultural biome (IAB), where researchers are studying the effects of high atmospheric carbon dioxide concentrations on food production.

to sprinklers in the roof to create high rainfall over the tropical rain forest. Mist is released periodically throughout the day from a network of pipes to form a cloud around the mountaintop. The water trickles back toward the marshes and the ocean through soil filters, providing a continuous supply of fresh water for both humans and ecosystems.

When first constructed in 1988, the Biosphere 2 project had lofty goals: to explore the scientific frontiers in eco-technology in order to promote better management of Earth's resources and to act as a model for colonizing space. Toward that end, between 1991 and 1993, eight "Biospherans" were voluntarily confined within the futuristic glass-and-steel greenhouse. Biosphere 2 was designed to be largely self-sustaining in terms of food and atmosphere. In theory, carbon dioxide from respiration is reabsorbed and the oxygen is replenished through photosynthesis. All wastes, including human and animal excrement, were to be treated, decomposed, and recycled back to support the growth of plants. Humans were to be nourished by animals, fish farming, and photosynthesis.

In practice, millions of dollars' worth of eco-technology could not maintain the atmosphere in optimum condition for life on Earth. In theory, photosynthesis was supposed to produce oxygen, but the double-paned glass cut 50 percent of the sunlight. The lack of ultraviolet light inhibited vitamin production and the ability of insects to find flowers to pollinate. Curing concrete and an oversupply of soil bacteria depleted oxygen levels: it was calculated that the enclosure lost fifty tons of oxygen, bottoming out at 14.5 percent (compared with Earth's 21 percent) after sixteen months of closure. Carbon dioxide levels, meanwhile, grew at times to 4,500 parts per million (current carbon dioxide levels in Earth's atmosphere are estimated at 360 parts per million). By the end of the two-year experiment, nitrous oxide levels were dangerously high, and contaminated water rained on the vegetation. The carbon dioxide dissolved in the ocean and formed carbonic acids, which began to dissolve and degrade the coral reef. Six hundred of the 3,800 species became extinct, including insect pollinators, requiring the hungry Biospherans to pollinate many plants by hand. A tiny Arizonan ant, *Paratrechina bruesii*, flourished, as did cockroaches. Partway through the experiment, supplements of oxygen were added, and ultimately 200,000 gallons of water and twelve tons of soil had to be replaced. (A subsequent six-month experiment was more successful; the crew had plenty to eat, insect pests were mostly under control, oxygen levels were up, and carbon dioxide levels were down.)

In 1994, the original managers of Biosphere 2 were ousted and replaced by a group of scientists led by Columbia University's Lamont Doherty Earth Observatory. Currently, Columbia has a five-year arrangement to manage and direct the biosphere's scientific and educational operations. Among areas of study are biodiversity and sustainable agriculture. Biosphere 2's high carbon dioxide levels make it an ideal place to study its effects. Researchers will assess whether increased $CO_2$ levels will allow farmers to grow more food with less water, and whether the increased plant growth created by extra carbon dioxide can significantly offset the negative aspects of global warming.

Meanwhile, observers of Biosphere 2's turbulent history contend they have learned a profound lesson about the limits of environmental technology. "As a species, we blithely take our planet's ecosystem services for granted," says conservation biologist John C. Avise. "The greatest lesson to come from Biosphere 2 should be a much greater appreciation for Biosphere 1, our own planet, which is also ultimately no less vulnerable to the impacts of human abuse.

"A more reliable path toward sustainability would seem to lie in using our intelligence and wisdom to understand and appreciate ecological principles and in using our technological skills to bring our human system into compliance with those principles."

# Conclusion

# *Great Expectations*

It wasn't supposed to have happened like this. The millennium, the target of so many futurists' imaginations, has turned out to be a bust. Predictions that were to have materialized—from space travel to Mars to nuclear mining of metals to the household robot—have been displaced by an obsession with the time setting on a computer clock. Even before the stroke of midnight on January 1, 2000, or a year later for millennial nitpickers, revisionist accounts of the future were being drafted. A signal event occurred in 1998, when Disneyland opened a refurbished Tomorrowland, a paean to the notion of unceasing progress. Gone was the People Mover. Disney originally wanted to market these monorail trains to transportation developers eager to speed commuters to work above congested urban streets. But developers did not line up for the technology the way that Disneyland patrons did for the original ride. In fact, the train consumed too much energy, even for a theme park ride.

The revamped exhibit was an attempt to recover lost innocence. Billed as "a classic future environment," its influences are Jules Verne, Leonardo da Vinci, and the now-quaint visions put forward at the 1939 World's Fair. Instead of a great big beautiful tomorrow shining at the end of every day (the motto of one now-defunct park exhibit), the new Tomorrowland is steeped in nostalgia for an earlier, more optimistic time— or at least one that was imagined as such. "The new flying rockets look more Flash Gordon than NASA," commented *Time* magazine. "The cars for a high-speed, roller-coaster-like attraction suggest a Victorian version of the Batmobile—H. G. Wells by way of Tim Burton."

The new Tomorrowland reflects the paradoxes of an age in which we embrace technology but pine for the past. The decoding of all human

285

genes proceeds apace even while we toast the "latest" automotive sensation: the Volkswagen Beetle. Our ambivalence may stem from the fact that futurism fascinates as much for its comical bad calls as for its pretensions to soothsaying. Any look at the state of what is to come, including the one found in this book, is bound to be at least partially wrong. Computers and biotechnology—the two paradigmatic technologies of the closing years of the twentieth century—have already fallen short of some of their seemingly endless promise. The birthday of HAL—the rogue computer of Arthur C. Clarke's *2001*—was January 12, 1997. That date, though, passed without so much as a peep from a real "thinking machine." Computers, moreover, may never match the ability of the human brain to cogitate. "Understanding how a single nerve cell works is an incredible challenge," notes John Guckenheimer, a professor of applied mathematics at Cornell University. "We don't really have a clue how neural systems function as reliably as they do. The kinds of digital computers that we build will certainly not develop the capacity to operate at the level of the human brain, because the number of interacting variables is so large that trying to simulate their behavior is completely beyond the bounds of what we are able to do."

Even more mundane applications of computers often disappoint. The paperless office is less a breakthrough technology of the turn of the century than a fondly remembered artifact of the heyday of the minicomputer in the 1970s. Computers, it turns out, demonstrate an ability to generate more hard-copy clutter than could the most industrious IBM Selectric typewriter.

Sometimes physics and chemistry simply refuse to cooperate with technologists' best-laid plans. If batteries for portable computers, cellular telephones, and the like had followed the same trajectory as the microchip, notes battery expert Isidor Buchmann, a car battery would now be the size of a coin, and cellular phone users would not kick themselves for forgetting to recharge their telephone batteries overnight. Similarly, Moore's law, the most oft-cited barometer of technological progress in the chip industry, may also be under siege. The process of patterning ever-thinner circuit lines atop a chip may be reaching its limits. Although a plethora of ideas have been put forward as replacements for conventional lithographic manufacturing methods, no one has yet shown that the alternatives will provide the ever-dropping cost spiral that has fueled the industry for so long.

Biotechnology has experienced an equally acute quotient of hyperbole. Predictions that the products of gene manipulation will become as important a force in shaping the modern world as the computer and its constituent microchip have yet to materialize. Fired by prospects of enormous profits for curing such intractable human problems as cancer, heart disease, and aging, venture capitalists have poured billions of dollars into biotech companies. Large pharmaceutical corporations have paid hefty premiums to purchase or invest in small biotechnology boutiques. But to date, many of the interested parties—researchers, investors, physicians, and patients—continue to bide their time. In many instances, drugs announced with great fanfare have turned out to be costly and sometimes dangerous failures. Gene therapy, the laborious attempts to correct defective DNA, is still a long way from producing cures for diseases like cystic fibrosis. The promise of new, targeted treatments for cancer beckons, but only after a series of early failures that could be repeated. Agricultural biotechnology has experienced setbacks and raised worries. A genetically engineered tomato did prove to be tastier than its pale, rock-hard, truck-ripened counterpart, but it did not travel well on its journey from field to supermarket and was summarily pulled from the shelves. And it is feared that crops genetically engineered to produce a natural insecticide might spur resistance by insects to the chemical.

So why do bad things happen to seemingly good ideas? The far-flung musings of the technological visionary have often derailed for simple reasons of cost and practicality. Take space travel. A mission to Mars has not been ruled out, and construction of the International Space Station has begun, but national or international resolve to undertake ambitious manned space sojourns has been sorely taxed. Perhaps the most enduring recent image of humans in space was that of Russian cosmonauts trying to salvage their decaying space station from a fire and other mishaps. The ill-fated experience of the inhabitants of Biosphere 2—pumping in oxygen to their earthbound desert abode—reveal the difficulties of maintaining any kind of permanent home separate from its environment.

One does not have to leave the Earth's atmosphere to experience the limits of technology. If Rip van Winkle had fallen asleep on a park bench in New York City in 1938 and awakened sixty years later, he would have found a city with some of its fundamental features relatively unchanged. The skyscrapers that have been an attribute of the Manhattan skyline since the early 1930s have not reached much higher than the Empire State Building, which was constructed in 1931. The subways that rumble beneath the city streets travel on rails that were first laid at the turn of the century. The potholed thoroughfares of Manhattan are paved with asphalt, a material first formulated before Christ lived. There is no technical reason why buildings cannot reach into the clouds. One could design and build Frank Lloyd Wright's mile-high edifice. But it would be an inherently wasteful endeavor. As buildings grow taller, elevator shafts take up an increasing amount of the space that a developer would prefer to allot to leasable offices. Sizable shafts are needed to fit the pulleys and counterweights. But an alternative method of propelling an elevator that is as energy-efficient as this 150-year-old technology has yet to be developed.

The absence of a pragmatic bent was not the futurists' only fault. These dreamers too often failed to ground grandiose imaginings within a societal context that gauged the environmental consequences of any new technology. The 1950s concept of using nuclear explosives to dig big ditches now seems surreal. Thus the road ahead may be less related to Jules Verne, H. G. Wells, and the Jetsons than to the contents of the environmental impact statement submitted by a builder to the Environmental Protection Agency.

Futurism also often errs because looking ahead leaves so little time for assessing the lessons of the past. Technological success requires translating basic science into things that work, as well as patience, commitment, and deep pockets. A country or region (such as Europe) must be rich enough to fund basic research into science and technology—investigations that are not directed toward the goal of producing a specific commercial product.

Evidence shows that most substantive technological advances come from fundamental research funded by public sources. A survey published in 1997 by CHI Research found that 73 percent of the science papers cited in recent U.S. patents made reference to research conducted by government or nonprofit organizations. Strategic research and development during the Cold War made possible solar power, lasers, satellites, and a number of communication technologies and advanced materials. But what role the government *should* play still remains a matter of heated debate. Some federal legislators point to the growing involvement of private industry in backing research and development. In 1997, U.S. industry laid out more than $130 billion in research and development dollars, compared to a figure of some $63 billion for the government. Moreover, the largest fraction of government funds was devoted to military research and development.

Still, most of the industrial money went to various forms of applied research, which is oriented toward development of a given technology, not toward the underlying scientific insights that lead to the final product. In basic research, by contrast, the government, with its roughly $18-billion contribution, remains the dominant funding source. But here, too, the picture is not clear. Despite much-publicized cutbacks in basic research at Bell Laboratories and other renowned institutions, the overall private outlay has grown to some $8 billion. Microsoft, for one, had plans to spend hundreds of millions of dollars over five years to expand investigations that are not oriented directly toward product development.

The nuances of arguments about basic research get overshadowed by the reality that pouring money into the laboratory is not enough to ensure leadership. National culture remains as important as research prowess. England, for one, owed its success during the Industrial Revolution to that country's strong entrepreneurial tradition. The essentially uneducated craftsmen who built the first machines of the Industrial Revolution found backing from bankers willing to support speculative undertakings. In France, although there was national support for applied science, the country's brilliant engineering scientists were handicapped because of a lack of a strong entrepreneurial culture. A more contemporary parallel can be found in the example of U.S. venture capital firms, which have played a leading role in the success of computer and biotechnology startups. And while Europe has established itself in particle physics and other areas of pure science, it has generally been more cautious in taking the risks needed to bring the cutting-edge technology from the laboratory to the marketplace.

To succeed, a good technology must be useful, and it must be able to survive the gauntlet of the marketplace, as well as economic and social conditions and government policies. But even those prerequisites may not suffice. Such unpredictable elements as timing, fashion, luck, and the quirks of human nature may prove equally important. There are also the vagaries of what is called the positive feedback cycle—in which one technology gains a slight advantage in the marketplace that, over time, allows it to gain a dominant position. At the dawn of the automotive age, for example, a steam-powered car set the world speed record of 122 miles per hour. Some of the success of the internal combustion engine could be attributed to Henry Ford's facility for devising a method of assembling his Model T so that middle-class Americans could afford it. But there was also dumb luck. An outbreak of hoof-and-mouth disease in 1914 closed public access to water for the boilers for steam vehicles. Thus, in this brief period, the gasoline engine triumphed—and essentially cut off other options. Indeed, one of the obstacles to current efforts to develop an emission-free car is the huge installed base of factories geared to produce hydrocarbon-powered internal combustion engines. Similar processes have been at work in recent years. For example, the VHS format for videocassette recorders and Microsoft's operating systems for personal computers triumphed, despite the availability of alternatives for both products that many engineers and consumers considered to be technically superior.

Building height may have remained more or less constant, and no travel agent sells tickets to Mars. But the rapid pace of change in other areas—with a new generation of microchip arriving every three years—has caused people to become somewhat jaded to the wonders of new technology. In some cases the notion of technological progress has produced its own backlash. A small neo-Luddite movement hails the joys of a simple existence carried on without the encumbrances of microwave ovens, cell phones, and personal computers. The dystopian world wrought by the worship of new technology runs as a thematic thread through movies like *Blade Runner* and *Jurassic Park*, and the "Dilbert" comic strip.

Nerd-bashing is offset, however, by the growing realization that the messes wrought by technology will likely be remedied only through the application of yet other technologies. Industrial ecology seeks to transform waste generated by manufacturers by turning it into new raw materials. Developing nations contemplate the possibility of leapfrogging the costly infrastructure of older technologies. Already the growing affordability of wireless and satellite communications enables access to phone service in areas where messages are still transmitted by walking to the next village. New solar power generators deliver electric power to villagers located hundreds of miles from the nearest electric substation. In the end, embracing the possibilities of applied physics, chemistry, and biology may be the only way that 6 billion humans and countless other creatures can continue to survive and coexist on planet Earth.

# Further Reading

Aldersey-Williams, Hugh. *The Most Beautiful Molecule: The Discovery of the Buckyball*. New York: John Wiley and Sons, 1995.

Aldridge, Susan. *The Thread of Life: The Story of Genes and Genetic Engineering*. London: Cambridge University Press, 1996.

Amato, Ivan. *Stuff: The Materials the World Is Made Of*. New York: Basic Books, 1997.

Ashall, Frank. *Remarkable Discoveries!* New York: Cambridge University Press, 1994.

Bains, William. *Biotechnology from A to Z*. New York: Oxford University Press, 1993.

Benyus, Janine M. *Biomimicry: Innovation Inspired by Nature*. New York: William Morrow, 1997.

Bray, John, *The Communications Miracle: The Telecommunications Pioneers from Morse to the Information Superhighway*. New York: Plenum, 1995.

Brennan, Richard P. *Levitating Trains and Kamikaze Genes: Technological Literacy for the Future*. New York: John Wiley and Sons, 1994.

Breslow, Ronald. *Chemistry Today and Tomorrow: The Central, Useful, and Creative Science*. Washington, D.C.: American Chemical Society, 1997.

Brody, David Eliot, and Brody, Arnold R. *The Science Class You Wish You Had: The Seven Greatest Scientific Discoveries and History and the People Who Made Them*. New York: Berkley, 1997.

Brown, Lester R., Michael Renner, and Christopher Flavin. *Vital Signs, 1998: The Environmental Trends that Are Shaping Our Future*. New York: Worldwatch Institute/W.W. Norton, 1998.

Buchmann, Isidor. *Batteries in a Portable World*. Vancouver: Cadex Electronics, 1997.

Cardwell, Donald. *The Norton History of Technology*. New York: W.W. Norton, 1995.

Cetron, Marvin, and Davies, Owen. *Probable Tomorrows: How Science and Technology Will Transform Our Lives in the Next Twenty Years*. New York: St. Martin's Press, 1997.

Denning, Peter J., ed. *Computers Under Attack: Intruders, Worms and Viruses*. New York: Addison Wesley, 1990.

Dertouzos, Michael. *What Will Be: How the New World of Information Will Change Our Lives*. New York: HarperCollins, 1997.

*The Dorling Kindersley Science Encyclopedia*. London: Dorling Kindersley, 1993.

Emsley, John. *Molecules at an Exhibition: Portraits of Intriguing Materials in Everyday Life*. New York: Oxford University Press, 1998.

Farkas, Daniel H. *DNA Simplified: The Hitchhiker's Guide to DNA*. Washington, D.C.: American Association for Clinical Chemistry, 1996.

Flavin, Christopher, and Nicholas Lenssen. *Power Surge: Guide to the Coming Energy Revolution*. New York: Worldwatch Institute/W.W. Norton, 1994.

Flowers, Charles. *A Science Odyssey: 100 Years of Discovery*. New York: William Morrow, 1998.

Gates, Bill. *The Road Ahead*. New York: Penguin, 1996.

Gordon, James Edward. *The Science of Structures and Materials*. New York: Scientific American Books/W.H. Freeman, 1988.

Grace, Eric. *Biotechnology Unzipped: Promises and Realities*. Washington, D.C.: Joseph Henry Press, 1997.

Graedel, Tom E., and Braden R. Allenby. *Industrial Ecology*. Englewood Cliffs, N.J.: Prentice-Hall, 1995.

Green, James Harry. *The Dow Jones-Irwin Handbook of Telecommunications*. Homewood, Ill.: Dow Jones-Irwin, 1986.

Green, John. *The New Age of Communications*. New York: Henry Holt, 1997.

Gribbin, John. *In Search of the Double Helix: Quantum Physics and Life*. New York: Bantam, 1985.

Hazen, Robert M., and James Trefil. *Science Matters: Achieving Scientific Literacy*. New York: Anchor Books, 1991.

Hazen, Robert, with Maxine Singer. *Why Aren't Black Holes Black? The Unanswered Questions at the Frontiers of Science*. New York: Anchor Books, 1997.

Hecht, Jeff. *Understanding Fiber Optics*. Indianapolis: Sams, 1993.

Hummel, Rolf E. *Understanding Materials Science*. New York: Springer-Verlag, 1998.

Karinch, Maryann: *Telemedicine: What the Future Holds When You're Ill*. Far Hills, N.J.: New Horizon, 1994.

Kevles, Bettyann Holtzmann. *Naked to the Bone: Medical Imaging in the Twentieth Century*. New Brunswick, N.J.: Rutgers University Press, 1996.

Kirby, Lorne T. *DNA Fingerprinting: An Introduction*. New York: W.H. Freeman, 1992.

Kornberg, Arthur. *The Golden Helix: Inside Biotech Ventures*. Sausalito, Calif.: University Science Books, 1995.

Langton, Christopher G., ed. *Artificial Life: An Overview*. Cambridge, Mass.: MIT Press, 1995.

Levy, Harlan. *And the Blood Cried Out: A Prosecutor's Spellbinding Account of the Power of DNA*. New York: Basic Books, 1996.

McNeil, Ian, ed. *An Encyclopaedia of the History of Technology*. London: Routledge, 1990.

*The Merck Manual of Medical Information: Home Edition*. Whitehouse Station, N.J.: Merck Research Laboratories, 1997.

Milburn, Gerard J. *Schrödinger's Machines: The Quantum Technology Reshaping Everyday Life*. New York: W. H. Freeman, 1996.

National Research Council. *Materials Science and Engineering for the 1990s*. Washington, D.C.: National Academy Press, 1989.

Nebel, Bernard J., and Richard Wright. *Environmental Science: The Way the World Works*. Upper Saddle River, N.J.: Prentice-Hall, 1996.

Nelkin, Dorothy, and Susan M. Lindee. *The DNA Mystique: The Gene as a Cultural Icon*. New York: W. H. Freeman, 1995.

Nowotny, Helga, and Ulrike Felt. *After the Breakthrough: The Emergence of High-Temperature Superconductivity as a Research Field*. New York: Cambridge University Press, 1997.

Office of Technology Assessment. *Advanced Materials by Design*. Washington, D.C.: United States Government Printing Office, 1988.

———. *Green Products by Design: Choices for a Cleaner Environment*. Washington, D.C.: United States Government Printing Office, 1992.

Owens, Frank J., and Charles P. Poole Jr. *The New Superconductors*. New York: Plenum, 1996.

*Popular Science*. New York, N.Y. Subscription inquiries: P.O. Box 51286, Boulder, CO 80322. Telephone: 1-800-289-9399.

Raeburn, Paul. *The Last Harvest: The Genetic Gamble That Threatens to Destroy American Agriculture*. New York: Simon and Schuster, 1995.

Rawlins, Gregory E. *Moths to the Flame: The Seductions of Computer Technology*. Cambridge, Mass.: MIT Press, 1996.

———. *Slaves of the Machine: The Quickening of Computer Technology*. Cambridge, Mass.: MIT Press, 1997.

Reader's Digest. *How in the World? A Fascinating Journey Through the World of Human Ingenuity*. Pleasantville, N.Y.: Reader's Digest Association, 1990.

Reid, T. R. *The Chip: How Two Americans Invented the Microchip and Launched a Revolution*. New York: Simon and Schuster, 1984.

Rifkin, Jeremy. *The Biotech Century: Harnessing the Gene and Remaking the World*. New York: Putnam, 1998.

Rissler, Jane, and Margaret Mellon. *The Ecological Risks of Engineered Crops*. Cambridge, Mass.: MIT Press, 1996.

Russo, Enzo, and David Cove. *Genetic Engineering: Dreams and Nightmares*. New York: W. H. Freeman, 1996.

Schechter, Bruce. *The Path of No Resistance: The Story of the Revolution in Superconductivity*. New York: Simon and Schuster, 1989.

*Science News: The Weekly Newsmagazine of Science*. Washington, D.C. Subscription department: P.O. Box 1925, Marion, Ohio 43305. Telephone: 1-800-247-2160.

*Scientific American*. Subscription inquiries: 800-333-1199 (United States and Canada); 515-247-7631 (other).

Shroyer, Jo Ann. *Quarks, Critters, and Chaos: What Science Terms Really Mean*. New York: Prentice-Hall, 1993.

Shurkin, Joel. *Engines of the Mind*. Rev. ed. New York: W.W. Norton, 1996.

Silver, Lee M. *Remaking Eden: Cloning and Beyond in a Brave New World*. New York: Avon, 1997.

Stallings, William. *Computer Organization and Architecture: Principles of Structure and Function*. 2nd ed. New York: Macmillan, 1990.

Tenner, Edward. *Why Things Bite Back: Technology and the Revenge of Unintended Consequences*. New York: Knopf, 1996.

Theodore, Mary K., and Louis Theodore. *Major Environmental Issues Facing the Twenty-first Century*. Upper Saddle River, N.J.: Prentice-Hall, 1996.

Trefil, James, and Robert M. Hazen. *The Sciences: An Integrated Approach*. New York: John Wiley and Sons, 1995.

Tudge, Colin. *The Engineer in the Garden: Genes and Genetics from the Idea of Heredity to the Creation of Life*. New York: Farrar, Straus and Giroux, 1993.

Turton, Richard. *The Quantum Dot: A Journey into the Future of Microelectronics*. Oxford University Press, 1995.

Whittle, David B. *Cyberspace: The Human Dimension*. New York: W.H. Freeman, 1997.

Wildavsky, Aaron. *But Is It TRUE? A Citizen's Guide to Environmental Health and Safety Issues*. Cambridge, Mass.: Harvard University Press, 1995.

Winner, Langdon. *The Whale and the Reactor: A Search for Limits in an Age of High Technology*. London, Ill.: University of Chicago Press, 1986.

Wyke, Alexandra. *Twenty-first Century Miracle Medicine: Robo Surgery, Wonder Cures, and the Quest for Immortality*. New York: Plenum, 1997.

Wynn, Charles M., and Arthur W. Wiggins. *The Five Biggest Ideas in Science*. New York: John Wiley and Sons, 1997.

# Index

acid rain, 262–64
actuator, defined, 221
ADA deficiency, 136, 147
Adleman, Leonard M., 27
advanced ceramics, 203–4
advanced composite materials, 201–2
aerodynamic design, 215–16
aging, defined, 221
agriculture, 139–41
agrobacterium, 141
AIDS and HIV virus, 119–20, 136
   blindness from, 152
   drug design and, 151
   genome, sequencing of, 146
   isolation of, 146
   research, mice, 147
airliners, crashes, 104, 139
air pollution, 260, 265–67
algorithm, 41, 52
alloy, defined, 221
Alzheimer's disease, 177
amorphous, defined, 221
amorphous silicon, 239
Ampere's Law, 209
analog signals, 55–56, 67, 87
analytical engine, 10, 14
angiogenesis, 158–59
angiography, 170
angiostatin, 158
antibodies, 161–62
antisense nucleotides, 151–53
Arber, Werner, 143
argon lasers, 92–93, 101, 102
arithmetic logic unit (ALU), 18
ARPANET, 73
arthroscopy, 180
artificial intelligence, 40–41, 52, 153
artificial life, 50–51
assemblers, 35
assembly language, 35, 52
atom, defined, 221
atomic force microscopes, 26
atomic-scale construction, 194–95
   materials design, 197–98
"atomic spray painting," 194
attenuated rays, 167
automobiles, 10, 247–51, 266–67, 278
Autonomous Land Vehicle (ALV), 46
avatar, 83
Avery, Oswald T., 109
Avise, John C., 284

Babbage, Charles, 10, 14
Bacillus thuringiensis (BT), 142
bacteria, 119–20
Baekeland, Leo, 218
Bain, Alexander, 85

bakelite, 218
Baltimore, David, 145
bandwidth, 56, 87
bar-coding, 96
base pairs, 3, 115–16
BASIC (computer language), 36, 37
batteries, 249–50
baud, defined, 87
Becquerel, Alexandre-Edmond, 240
Bell, Alexander Graham, 56, 85
Bell Laboratories, 37, 98, 240, 288
Bhopal, India, 275
binary/base 2, 31
binary logic, 12–14
binary numbers, 13–14
Binnig, Gerd, 26
bioaccumulation, 268–69
biodigesters, 235
biomagnification, 268–69
biomimetic materials research, 219–20
bionics, 186–88
bioremediation, 4, 271–73
Biosphere 2, 282–84, 287
biotechnology, timeline, 145–48
birth defects, 158
bit, defined, 31, 87
blood oxygen level dependent (BOLD) imaging, 174
BodyNet, 82–83
Boole, George, 13
Boolean algebra, 13, 31
borrelia burgdorferi, 148
Bose-Einstein condensation, 105, 221
brassica (mustards), 273
breast cancer gene (BRCA), 130, 132, 147, 148
Brenner, Sydney, 110, 115
Breslow, Ronald, 254
bromoxynil, 147
bronchoscopy, 180
Brooks, Rodney A., 45
buckminsterfullerene ($C_{60}$), 211–14
buckyballs, 211–14
bunnyballs, 213
bus, defined, 31
byte, defined, 31

C (computer language), 37, 38
caches, CPUs, 21–22, 32
Caenorhabditis elegans, 148
calcium sulfate ($CaSO_4$), 265
cancer research, 130–31, 154–63
carbon, 196
carbon dioxide, 225–33
carbon energy economy, 224–25, 232–33
carrier signal, defined, 87
catalysts, 259
cavitron ultrasonic surgical aspirator (CUSA), 178

CD players, lasers, 96–97
cellular communications, 61, 64–67
central office, defined, 87
central processing unit (CPU), 17–23, 32
ceramic matrix composites (CMCs), 201
cermets, 203, 221
CERN High Energy Physics Laboratory, 75
chemokines, 151
chemotherapy, 156
Chernobyl, 244
chirality, 196
CHI Research, 288
chlorofluorocarbons (CFCs), 257–62
chromosomes, 126
cladding, 60
Clarke, Arthur C., 44, 45, 62, 286
Clean Air Act amendments of 1970, 260, 264–66
Clean Water Act of 1972, 267
Clinton, Bill, 84
clock speed, 32
cloning, 120–25
    humans, 124–25
    mammal cells, 108, 121–24
    mice, 123–24
    microorganisms, 121
    plants, 121, 122
    sheep, 108, 122–24
COBOL (computer language), 36, 37
Code Division Multiple Access, 67
coding, 34
codons, 112
Cog (robot), 45
cogeneration, 234
Cogen Technologies, 234
coherence, defined, 106
Cold Spring, N.Y., 269–70
Cold War, 7, 11, 288
combinatorial chemistry, 153–54, 197
communication pipes, 58–61
compiler, software, 52
complementary DNA (cDNA), 117
complex instruction set computing (CISC), 22, 32
composite, 221
compounds, 221
compressive strength, 221
computational materials research, 197
computer assisted tomography (CT), 166, 168–71
computer basics, 9–32
computerized axial tomography (CAT), 169
computerized digital-scan converters, 177
conductor, 221
conservation reserve, 233
controller, CPU, 32
Cooper pairs, 207
copper wiring, communications, 58–59
corn, genetically altered, 143
Crick, Francis, 5, 107, 109–10, 115
criminal investigations, DNA, 137–39
crystal, 221
Curie, Marie and Pierre, 199
Cuyahoga River fire, 256
CYC project (CYClopedia), 44

cystic fibrosis, 136
cytokines, 162

Darwin, Charles, 109
data compression, defined, 87
Data Encryption Standard, 80
data networks, 53–88
Deep Blue, 42
deep ecology, 274
Defense Advanced Research Projects Agency
    (DARPA), 185
Department of Defense, 67
Dernococcus radiocurans, 148
design for disassembly, 274
design for the environment (DFE), 274
diagnostics, medical, 165–79
digital communications, 55–57
digital information, lasers and, 96–97
Digital Subscriber Line (DSL), 56
disk drive, 17
Disneyland, 285
DNA (deoxyribonucleic acid), 107–48, 287
    computing and, 26–29
    cutting, pasting, and copying, 6
    delivery methods, 136–37
    discovery of, 5
    establishment of structure, 107
    features, 117
    Human Genome Project. See Human
        Genome Project
    milestones, early, 109–10
    transcription (messenger RNA), 113–14
DNA ligase, 120
doping, 15, 221
DRAMS, 3, 24
drug research, 149–63
Duchenne's muscular dystrophy, 136, 146, 157
duplexing, defined, 87
DVD players, 96–97
dye lasers, 94

eco-development, 274
Edison, Thomas Alva, 85
Einstein, Albert, 89, 90
electromagnetic spectrum, 57, 87
electrophoresis, 129
electrorheological fluid, 199
electrostriction, 200
ELIZA (shrink-in-a-box program), 44
E-mail, 54, 55, 74–75
encryption, 78–80
endoscopy, 180–81
endostatin, 158
energy, 223–54
ENIAC, 11–12
environment, 6–7, 133–34, 255–85
environmental remediation, 269–73
enzymes, 107, 111, 114, 116
epitaxy, 221
erbium-doped fiber amplifiers, 61
Escherichia coli (E. coli), 112–13, 218, 275
ESS (Electronic Switching System), 70, 87

ethernet, 69
Everest Extreme Expedition, 184
excimer lasers, 94, 103
expert systems, computers, 41, 52
expressed sequence tags, 117
extracellular matrix, 190
extrude, 221
Exxon Valdez oil spill, 272

Farkas, Daniel H., 115, 118
farnesyl transferase, 157
feedback, global warming and, 228–29
fiber optics, 58–61, 87, 179–81
field programmable gate arrays (FPGAs), 23
Fifth Generation Initiative (Japan), 43
firewalls, 79
Flemming, Walther, 109
flywheels, 250–51
Folkman, Judah, 163
Ford, Henry, 289
FORTRAN (computer language), 36, 37
fossil fuels, 224, 225
free-electron laser (FEL), 94, 179
Frequency Division Multiple Access, 65
Frequency division multiplexing, 87
frequency modulation, 87
fuel cells, 251–52
fullerene chemistry, 211–14
functional MRI (fMR), 174
futurism, 288
fuzzy logic, 48–49, 52, 153

games, computers, 42–43
Gates, Bill, 9, 34, 36, 64, 83
gene chips, 117, 155
gene library, 122
Genentech, Inc., 146
General Circulation Models, 228
gene therapy, 135–37, 287
genetic engineering, 5–6, 113, 139–45
geo-engineering, 232
geosynchronous (geostationary) satellites, 63, 64
geothermal power, 241–42
Gilbert, Walter, 129, 145
Gilmartin, Raymond V., 150
Global Positioning System (GPS), 67–68
global warming, 4, 224–33
gold sulfides, 275
Gore-Tex, 252
Gould, Stephen Jay, 51
graphical user interface (GUI), 39
graphite, 196, 221
green design, 274
Greenfreeze, 261
greenhouse effect, 225–27
green products, 274
groundwater contamination, 270, 272
growth factors, 158
Guckenheimer, John, 286
Gulf War, 49, 68, 103–4

hackers, 78–80
hardware, computer, 4, 34

Haseltine, William, 128–29
HAZMATS, 268
health sciences, 6, 165–90
heavy-metal recycling, 277
helium neon lasers, 95
Helminthosporium maydis, 143
hemoglobin, 111
hertz, defined, 106
heuristics, 41–43, 52
high-temperature superconductors, 206–10
high-throughput screening, 153
Hocking, Martin, 281
hole, defined, 222
holograms, 98–99, 106
holographic storage, computers, 24–26
homunculus, 109
Hook, Robert, 109
Human Genome Project, 3, 5–6, 125–33, 147
human genome research, 133–35
Huntington's chorea, 146
hydroelectric power, 241–42
hydrogen, 252–53
hydrogen bomb, 11
Hypertext markup language (HTML), 75
hysteroscopy, 180

IBM, 35, 42, 260–61
IBM PC, 21, 22
icons, computers, 39–40
imaging, medical, 166–79
incoherence, 91
industrial ecology, 256, 274, 279–80
industrial symbiosis, 274
inference engine, 41
information age, health care and, 181–86
"information superhighway," 54
INMARSAT, 184
in situ hybridization, 146
insulator, 15–16, 222
insulin, production of, 113, 145
integrated circuits, 15–16, 32
Integrated Services Digital Network (ISDN), 56, 87
Internet, 54–55, 72–84
  e-mail. See E-mail
  World Wide Web. See World Wide Web
Internet Engineering Task Force (IETF), 72
Internet Protocol (IP) address, 74
Internet Society, 72
interoperative MRI, 174
interpreter, software, 52
ion, defined, 222
Iridium satellite network, 66, 86
isomers, defined, 222
isotopes, defined, 222

Jacquard, Joseph-Marie, 10
James Bay project, 241
Japan, Fifth Generation Initiative, 43
Java (computer language), 37
jet engines, 223, 235–36
Josephson, Brian, 210
Josephson junctions, 210–11, 222

junk DNA, 128–29

Kalundborg (Denmark), 279–80
Kasparov, Garry, 42
knowledge engineers, 42
Koestler, Arthur, 2
Kornberg, Arthur, 108

landfills, 276
languages, computer, 37–38
laparoscopy, 180
Laser Geodynamic Satellite, 95
lasers, 89–106
leaching, 270
lead-acid batteries, 249–50
Leder, Philip, 147
Leeuwenhoek, Antonie van, 109
Levy, Harlan, 139
life cycle analysis (LFA), 274, 280–81
Life Support for Trauma and Transfer, 183
limestone (CaCO₃), 263
Lincoln, Abraham, 133
liposomes, 136, 217
liquid nitrogen, 206
LISP (computer language), 41
local area networks (LAN), 69
logic gate, defined, 32
low-earth orbit satellites, 64
Lucent Technologies, 11, 60, 70, 98
Lyme disease, 148

machine code, 34–36, 52
Macintosh, 22
magnetic resonance elastography, 178
magnetic resonance imaging (MRI), 6, 166,
    171–74
magneto-encephalography (MEG), 178
Maiman, Theodore, 91, 92
Marconi, Guglielmo, 64, 85
Marfan's syndrome, 133
masers, 91, 93, 106
materials science, 6, 191–222
Matthaei, Johann, 112
Maxam, Allan, 129, 145
Mayer, Carol, 167–68
McCarthy, John, 41
McCune, William, 48
Medicams, 181
medium earth orbit satellites, 64
Meissner, Alexander, 210
memory, computer basics, 17–19, 24, 32
Mendel, Gregor, 109
Mendeleev, Dimitri, 193
Mendelian Inheritance in Man, 147–48
Merck & Co., Inc., 150
messenger RNA (mRNA), 113
metal matrix composites (MMCs), 201, 203
methane, 234–35
methyl isocyanate (MIC), 275
metropolitan area network (MAN), 69
mice
    cancer research, 158–59, 163

cloning, 123
    transgenic ("Harvard mouse"), 147
microchips, 9–11, 19–21
microelectromechanical systems (MEMs), 214
micromechanics, 214–16
micrometastases, 155
microprocessor, 20
Microsoft, 9, 34, 83, 289
Microsoft Windows, 39
microwaves, 87, 209
Midgeley, Thomas Jr., 260
Miescher, Friedrich, 109
military, materials science and, 193–94
minisatellites, 138
mitosis, 110
mobile communications, 61
mobile sources of air pollution, 266–67
modulation, 88
molecular-beam epitaxy (MBE), 98, 194–95
molecular biology, 6, 149–63
molecular computing, 26–29
molecules, 222
Molina, Mario, 260
Monju fast-breeder nuclear plant, 245
monochromatic light, 106
monoclonal antibodies ("magic bullets"), 160–62
Montreal Protocol on Substances that Deplete
    the Ozone Layer, 260–61
Moore, Gordon, 19
Moore's Law, 19–20
Morse, Samuel F. B., 85
Mount Pinatubo eruption, 228
MS DOS, 36
Mullis, Kary, 118, 146, 148
multimedia, 88

naphthalene, 273
NAP (network access point), 74
National Institutes of Health, 127, 146
National Priorities List, 268
National Science Foundation, 73
NAVSTAR Global Positioning System, 67–68
Nd:YAG laser, 93, 101
Nebel, Bernard J., 231
Nelson, Theodore, 86
networks, computer, 53–88
network topology, 70
neural networks, 46–47, 52
Newton, Isaac, 29
Nirenberg, Marshall, 112
nitric acid (HNO₃), 262
Nobel Prize, 108, 118, 143, 145, 148
nuclear medicine, 160–62, 175–77
nuclear power, 242–47
nuclear waste, 244–47
"numbers problem," 15

object beam, 99
object-oriented programming, 37–38, 52
Ogawa, Seiji, 174
oligonucleotides, 116
Olson, Maynard, 147

oncogenes, 156
Oncomouse, 123
Onnes, Heike Kamerlingh, 207
operating system, computers, 36–37, 52
ophthalmology, lasers used in, 101–3
optical fibers, 58–61, 87, 97, 102
optical maser, 91
optical molasses, 104–5, 106
optics, defined, 106
*Origin of the Species* (Darwin), 109
Otto, Nicholas August, 247
ozone hole, 259–61
ozone layer, 258–62

packet switching, 88
paper cups, 280–81
parallel processing, 22–23, 32
Pascal, Blaise, 10
PASCAL (computer language), 37, 38
Pentagon, 43, 46
periodic table of the elements, 192–93
phospholipids, 217
photochemical oxidants, 259
photochemical smog, 259
photodynamic therapy (PDT), 159–60
photoelectric effect, 240
photolithography, 15–17, 32
photon, defined, 106
photorefractive keratectomy (PRK), 103, 106
photovoltaic cells, 239–40
phytoremediation, 273
piezoelectric materials, 199–200, 222
pipelining, CPUs, 21, 32
Planck, Max, 90
plastic, 191
pollution prevention, 234, 264
polychlorinated biphenyls (PCBs), 257
polymerase chain reaction (PCR), 117–18, 130,
    137, 139, 146, 148
polymer electrolyte membrane (PEM) cells, 252
polymer matrix composites (PMCs), 201
polymers, 190, 218, 222
polyvinylidene fluoride (PVDF), 200
population inversion, 106
positron emission tomography (PET), 6, 166,
    175–77
power towers, 240–41
protein kinases, 157–58
protocols, defined, 88
pseudomonas bacteria, 271
public-key cryptography, 79
pumping, defined, 106
punctuated equilibrium, 51
purines, 110
pyrimidines, 110

quantum cascade lasers, 98
quantum parallelism, 30
quantum theory, 29–30, 90
quantum well, 96
quasicrystals, 204, 206
qubits, 30

Rabi, I. I., 172
radar waves, 209
random access memory (RAM), 24
rare earth, 222
*Ras* genes, 157
Rawlings, Gregory E., 5
RBST (recombinant bovine somatotropin), 142
Reagan, Ronald, 90
recycling, 276–78
reduced instruction set computing (RISC), 22, 32
reference beam, 99
reflective light, 99
register, in computers, 32
"reporter genes," 273
reprogrammable chips, computers, 23–24
restriction enzymes, 116
restriction fragment length polymorphisms
    (RFLPs), 146
retroviruses, 119–20, 137
reverse transcriptase, 119
rinderpest virus, 147
RNA polymerase, 113
RNA (ribonucleic acid), 109–10, 119
ROBODOC, 169
robotics, 31
Roentgen, Wilhelm Conrad, 166
Rogers, Craig A., 201
Rohrer, Heinrich, 26
Rosen, Benjamin and Harold, 250–51
routers, 71, 88
Rowland, F. Sherwood, 260
RSA, 79
ruby laser, *91*, 92

Samuel, Arthur, 42, 51
Sanger, Fred, 129, 130
satellite communications, 61–64, 86
scanning probe microscope, 26
scanning tunneling microscope, 26, 198
Secure Electronic Transaction Protocol (SET), 79
security, data networks, 78–80
self-assembling monolayers (SAMs), 217
self-assembly, 216–17
semiconductor lasers, 96, 97
semiconductors, 15, 32
sense strand, 114
shape memory alloys, 200
sheep, cloning of, 108, 122–24
Shoreham nuclear power plant, 242–43
silicon, 191, 205–6
silicon chips, 9
Simpson, O.J., 139
single gene diseases, 130
single photon emission computed tomography
    (SPECT), 177
"smart" cards, 181, 205
"smart" eyeglasses, 82
"smart" materials, 198–201
"smart" underwear, 83
software, computer, 33–52
solar energy, 237–41

solar troughs, 240–41
Southern, Edward, 145
space travel, 287
spatial light modulator (SLM), 25
spectroscopy, 197
Spheres of Position, 68
spiral scanning, 169–71
spontaneous emission, 90, 106
Sputnik 1, 86
"Star Wars" (Strategic Defense Initiative), 90
steam engine, 256
Stewart Timothy, 147
"sticky ends," 116
stimulated emission, 90–91, 106
stratosphere, 258
structured programming, computers, 38, 52
subroutines, computer programs, 38
sulfur dioxide, 265
sulfuric acid ($H_2SO_4$), 262
superconducting quantum inference devices
    (SQUIDS), 178, 208–9, 210
superconductivity, 222
Superfund, 267–70
superscaling, 21, 32
surface micromachining, 215
susceptibility genes, 133
sustainable development, 274
switching calls, 70–72
systems programming, 36

talking computers, 43–44
Taq Polymerase, 118
Tay-Sachs disease, 131, 135, 136
technology
    "green," 273–76
    range of, 4
telecommunications, 53–88
Telecommunications Reform Act of 1996, 86
Teledesic project, 64
telehealth, 182
telemedicine, 6, 181–86
telementoring, 182
telepathology, 182
telepresence surgery, 182
telesurgery, 185
Telstar 1, 62, 86
Temin, Howard, 145
tempering, 222
tensile strength, 222
tevatron accelerator, 209
thermodynamics, 234
Thermus aquaticus, 118
thin-film transistors, 205
tidal power, 241–42
time division multiplexing, 88
tissue engineering, 6, 188–90
tomography, 166–68
Tomorrowland, 285
totipotent cells, 121
Townes, Charles H., 91
toxic dumps, 267–68

transcription, 114
transformation, 119
transistors, 11, 15, 32
Transmission Control Protocol, 74
transmissive light, 99
Trichloroethylene (TCE), 272
triplets, 112
troposphere, 258–59
tumor suppressor genes, 156
turbogenerators, gas, 235–36
Turing, Alan, 43
Turing Test, 44
Tuskegee University, 141
twos complement, 14

ultrasound, 177–78
ultraviolet light, 15–17
ultraviolet (UV) radiation, 258–59
unintended consequences, genetic engineering,
    142, 144–45
Union of Concerned Scientists, 140
United States Supermarket Institute, 96
Unix-to-Unix CoPy (UUCP), 75

vacuum tubes, 10–11, 14–15
Vail, Theodore, 85
variable number tandem repeats, 138
Venter, Craig, 127–28
Verne, Jules, 10, 285, 288
Vinci, Leonardo da, 285
virtual reality (software), 49–50
viruses
    biological, 119–20
    computer, 80
volatile organic compounds (VOCs), 224, 247,
    259, 264

The War of the Worlds (Wells), 89, 103
Waste Isolation Pilot Plant, 247
water molecules, 172
Watson, James, 5, 107, 109
wavelength, 88
wave properties, defined, 106
Weizenbaum, Joseph, 44
Wells, H. G., 89, 103, 285, 288
wide-area network (WAN), 69
wind turbines, 236–37
Wirth, Niklaus, 37
wireless communications, 61, 64–67
World Wide Web, 54, 75–77, 81, 86
World Wide Web Consortium (W3C), 73
Wright, Frank Lloyd, 287
Wright, Richard T., 231

X-ray holograms, 99
X-ray lasers, 94
X rays, 166, 167

yeast artificial chromosomes, 147

zero-emission vehicles, 247–49